DESIGN FOR AT-SPEED TEST,

DIAGNOSIS AND MEASUREMENT

TRADEMARK INFORMATION

DESIGN FOR AT-SPEED TEST,

DIAGNOSIS AND MEASUREMENT

Edited by

Benoit Nadeau-Dostie
LogicVision, Inc.

KLUWER ACADEMIC PUBLISHERS
Boston / Dordrecht / London

Distributors for North, Central and South America:
Kluwer Academic Publishers
101 Philip Drive
Assinippi Park
Norwell, Massachusetts 02061 USA
Telephone (781) 871-6600
Fax (781) 871-6528
E-Mail <kluwer@wkap.com>

Distributors for all other countries:
Kluwer Academic Publishers Group
Distribution Centre
Post Office Box 322
3300 AH Dordrecht, THE NETHERLANDS
Telephone 31 78 6392 392
Fax 31 78 6546 474
          E-Mail <orderdept@wkap.nl>

 Electronic Services <http://www.wkap.nl>

Library of Congress Cataloging-in-Publication Data
Design for at-speed test, diagnosis, and measurement / edited by Benoit
Nadeau-Dostie.
          p.   cm. -- (Frontiers in electronic testing)
   Includes bibliographical references.
   ISBN 0-7923-8669-8 (alk. paper)
      1. Integrated circuits--Testing. 2. Electronic apparatus and appliances--
Testing. I. Series. II. Nadeau-Dostie, Benoit.

TK7874.D47497 2000
621.385 21--dc21   99-046022

Printed on acid-free paper. Printed in the United States of America

Table of Contents

Foreword

System-on-a-chip (SOC) technology has come to represent various aspects of integrating more and more functionality in a single integrated circuit (IC) and delivering it to the market in the shortest possible time. This ranges from ICs with user-defined logic and embedded memories on one extreme end of an SOC, to a complex graphics controller SOC with an embedded microprocessor, several embedded memories, phase-locked-loops, analog-to-digital converters, digital-to-analog converters, user-defined logic, and other predesigned cores. Despite disagreements on a commonly accepted definition of an SOC, there is a widespread agreement that SOCs represent a major discontinuity with incredible design and test challenges. Now imagine a product built with several such SOCs. Most data communication products are perfect examples of such complex systems.

Many of these design challenges are being met by new hierarchical, block-based design methodologies and corresponding automation tools that include higher-level floorplanners, verification environments, static-timing signoffs, and block-based synthesis. Nonetheless, one of the key problems faced by designers is the manufacturing test, diagnosis and measurement. This continues to be perceived as a back-end problem and is ignored by most electronic design automation (EDA) tool providers. In this age of SOC, the test problem impacts time to market, quality, and overall cost.

What is needed is a total design for testability solution that can provide at-speed test of all embedded blocks and of the chip as a whole. This solution should be able to handle diversity of embedded cores from logic to memory to mixed-signal and their associated fault models. At the chip level, the solution should be able to reduce the test time, provide higher fault coverage, better diagnostics, and reduce the time to market. At the system level, fast time to market requires better tools to perform board-level prototyping and system-level integration. Reliability at a lower cost is also a prime objective for many applications.

It is well known that some of these problems are fundamentally impossible to solve with external testers. It is also generally accepted that a more economical approach to solving these problems is to provide a proper balance between external test and embedded test. Incorporating some of the test functions right on the very chip that is being tested makes an SOC include the tester as a part of that system. Embedded test—if done properly—offers a whole range of advantages for solving test, diagnosis, and debug problems.

This is the first book of its kind to offer practical and proven solutions to product managers, designers, and test engineers for test and diagnosis problems at the silicon level as well as the board and system levels. The basic theme of this book is that embedded test offers significant reductions in design and test engineering efforts, in time to market, and in cost of design and test of chips, boards, and systems.

LogicVision has been at the forefront of the embedded test technology with its continued innovations, automation, and customer successes. This book is an admirable effort by Dr. Benoit Nadeau-Dostie and his colleagues to share the fruits of their success with the design and test community at large. The practical aspects of design for testability explored in this book bring immediate value to its readers. System designers will find the material extremely useful in defining requirements for their chips that can speed up board-level prototyping, and system-level integration and bring up. Embedded test has the added benefit of reducing the extent of diagnostic software that must go on every complex system.

For chip designers, the use of embedded test significantly helps them meet their 99 percent plus fault coverage requirements with small test time and at-speed tests. In addition, designers benefit from help of embedded test in performing silicon debug of their first samples. Finally, the product engineers and test engineers can take full advantage of embedded test by reducing their efforts in the preparation of complex test programs and diagnostics.

Prem Jain
Vice President and General Manager
Enterprise Line of Business
Cisco Systems, Inc.
San Jose, CA.

Acknowledgments

Many people contributed to the writing of this book, which required a considerable team effort. I thank all the contributors for their sustained enthusiasm and diligence.

A large team of highly competent engineers was involved in the development and application of the products and capabilities referred to in this book. I gratefully acknowledge all of them for making the embedded test technology robust and easy to use.

We are deeply indebted to all our loyal customers, who have given us the chance to prove and improve the technology. In particular, some companies made special contributions. Nortel provided access to their memory and logic BIST technologies, which constituted the first generation of LogicVision products. Stratus Computers helped us specify and use the first controllers to support multi-cycle paths—a significant milestone in at-speed logic testing—as well as the first controllers to support at-speed interconnect testing. Fujitsu was the main driver and first user of embedded core test technology. Lockheed-Martin contributed to the development of off-the-shelf memory test technology in the context of the US Air Force Rapid Prototyping Program (RASSP). Mitel was the first user of the new embedded test technology developed for analog-to-digital converters. Ericsson provided the motivation for developing the fault insertion feature.

Dr. Vishawani D. Agrawal, Series Editor of Kluwer Academic Publishers, was very helpful in promoting the concept of this book, and I wish to thank him for his support. Carl Harris at Kluwer Academic Publishers also deserves our gratitude. Finally, special thanks to Susan Borton and Diane Lynch for converting the manuscript into a truly professional publication.

About the Editor

Benoit Nadeau-Dostie has been Chief Scientist at LogicVision since 1994, defining test strategies and algorithms with a focus on embedded test. He obtained his Ph.D. degree in Electrical Engineering from Université de Sherbrooke, Québec, Canada. From 1986 to 1994, he was an Advisory Engineer at Bell-Northern Research (BNR) and was the main architect of their Design-for-Testability (DFT) strategy. From 1985 to 1986, he was a member of the Department of Electrical Engineering of Université Laval (Vision and Digital Systems Lab). He has published several articles and holds 10 U.S. patents in the areas of memory, logic, and board testing. Dr. Nadeau-Dostie is a senior member of the IEEE and is involved in many of its activities, including international conference program committees and the IEEE 1149.1 Standard Working Group. He is a member of the Editorial Board of *IEEE Design & Test of Computers*.

About the Contributors

The following contributors participated in writing this book.

Stephen Baird is a marketing director at LogicVision with responsibilities for product and account development. Prior to joining LogicVision in 1997, he worked for Texas Instruments for 30 years in various roles including wafer fab process engineering, ASIC applications and product development, and IC design and product engineering. He received a BSEE from Southern Methodist University.

Martin Bell is a lead software engineer with 11 years of experience automating a wide variety of DFT methodologies. Prior to joining LogicVision, he worked at VLSI Technology and its subsidiary, COMPASS Design Automation, where he developed the COMPASS Test Assistant. Mr. Bell has published DFT papers at ITC and GOMAC, and is the inventor of two DFT patents. He received his MSEE from Stanford University.

Dwayne Burek received an MSEE from the University of Manitoba. Prior to joining LogicVision he was a Member of the Scientific Staff at Bell-Northern Research (BNR), which is now Nortel Networks. At BNR he was responsible for defining, developing, and deploying DFT solutions across a wide portfolio of products. He is a member of the IEEE P1500 Architecture Task Force and the Virtual Socket Interface Alliance (VSIA) Manufacturing-Related Test Development Working Group.

Jean-François Côté is a Design-for-Reliability expert with 10 years of experience in design. He designed several dynamic RAMs at Mosaid. He was the ASIC architect of Nortel's OC-192 optical transport equipment. He has an MSEE from McGill University and holds four U.S. patents.

Harry Hulvershorn has 12 years experience in system DFT and BIST, specializing in board and system test. Prior to joining LogicVision, he worked as a Test Engineer at Unisys Defense Systems, and later as a System DFT advisor at Nortel. He chairs the IEEE 1149.5 Module Test and Maintenance Bus Working Group. He has a BSEE from the University of Manitoba.

Diane Lynch is a publications manager with over 13 years of experience in designing and authoring first-generation technical documentation on database and library facilities, analog and digital simulation, PCB design and layout, and DFT. Prior to joining LogicVision, she was employed by Mentor Graphics, Daisy/Cadnetix (DAZIX), and IBM. She graduated cum laude with a BA in English from Rhode Island College.

Fadi Maamari has been with LogicVision since 1996, where he is currently Director of the Design Automation group. Prior to joining LogicVision he was with AT&T Bell Labs, where he worked on DFT, Fault Simulation, and ATPG. He received his Ph.D. in Electrical Engineering from McGill University.

Stephen Pateras received his Ph.D. in Electrical Engineering from McGill University. Before joining LogicVision, he was a member of IBM's New System Technology and Architecture Division, where he led a team responsible for all test and DFT issues for the company's CMOS 5390 mainframe design project. He is on the program committee of the International Test Synthesis Workshop. He has authored and coauthored several papers and articles in the fields of test generation and BIST.

Stephen Sunter has over 20 years of experience in mixed-signal IC design, DFT, and test. He has held various positions at Bell-Northern Research (BNR) and Northern Telecom (now Nortel Networks) as a design engineer, design manager, and test manager. He has published numerous articles and holds seven U.S. patents. He has been Vice-Chair of the IEEE P1149.4 Working Group since 1994.

Yervant Zorian received a Ph.D. in Electrical Engineering from McGill University. Prior to joining LogicVision, he was a Distinguished Member of Technical Staff at Bell Labs, Lucent Technologies, and was responsible for developing embedded test strategies. A Fellow of the IEEE, he chairs the IEEE Test Technology Technical Committee (TTTC) and the IEEE P1500 Working Group. Dr. Zorian is the current Editor in Chief of the magazine *IEEE Design & Test of Computers*. He holds several patents and has authored more than 100 publications, including two books.

Preface

This book addresses designers, test engineers, and product managers, for whom success depends upon the fast deployment in the market of electronic systems based on complex ICs. The *system* can be the integrated circuit itself (system on a chip) or an arrangement of multiple boards. The complexity of today's chips is not only a function of the number of transistors but also the diversity of embedded functions such as logic, memory, processors, and mixed-signal components.

Product managers will read how the time, resources, and costs associated with test development, manufacture, and life-cycle maintenance of their products can be significantly reduced by designing test, diagnosis, and measurement features in the product. This book provides a complete design flow and analysis of the impact on the design to enable designers to learn how to implement these features in systems. Test engineers will learn how embedded test circuitry can provide superior quality without exceeding the capabilities of their equipment. The book also provides procedures for diagnosing failures and measuring circuit parameters.

The book places special emphasis on at-speed testing. Being able to perform tests, diagnose failures, and make measurements using system clocks offers several benefits. The main benefit is the ability to test the circuit—or subcircuit—under conditions that are as close as possible to normal circuit operation. This factor has a direct impact on the number of chips that are found defective during system operation but still pass all manufacturing and functional tests. At-speed testing can be used for characterization and can also expedite test application time.

Finally, the book addresses the concerns associated with embedded test. One such concern is the perceived inability of embedded test to perform proper diagnosis and the assertion that the technology can be used only to determine whether circuits pass or fail a test. Not only do the technologies described in this book have several levels of diagnostic features, but they can also perform accurate measurements of circuit parameters. Other concern about embedded test relates to the size of the test circuitry and its power consumption. It will be seen that both of these concerns are not significant with the solution described in this book.

The type of information in this book is significantly different from the information provided in other books on Design for Test (DFT) and Built-in Self-Test (BIST), which are fundamental components of embedded test. These books typically present the basic concepts of several potential solutions to the same test problem without providing all the information required to determine the application context and implementation issues. This book emphasizes the issues that the practitioner, designer, or test engineer must understand in order to implement embedded test for a single chip or an entire system. For example, the book addresses the following questions: What are the restrictions put on the design style of the chip and the system? Is there any impact on circuit performance? How are clock skew and other timing issues addressed? How are the tests verified and debugged?

All the technology described in this book has been proven in silicon, is automated, and is integrated with all the other components described. This technology has been used by hundreds of designers and test engineers in major companies, including Cabletron, Cisco Systems, Ericsson, Fujitsu, Hughes, Lockheed-Martin, LSI Logic, National Semiconductor, Nortel Networks, Stratus, and Sun Microsystems. All this is to underscore that everything you read in this book is practically sound and verifiable on real-world chips.

Overview of the Chapters

This book is organized as follows:

Chapter 1, "Technology Overview," has three main sections. The first section discusses the motivation behind, need for, and advantages of designed-in, or embedded, test. The second section introduces basic concepts of LogicVision's embedded test solution. The third section provides an overview of all embedded test capabilities.

Chapter 2, "Memory Test and Diagnosis," describes embedded test technology for memories. Memory BIST is a well-known application and is used to illustrate many of the features common to all LogicVision technologies. For example, all controllers use a common protocol for initialization and results collection. Diagnosis and integration with scan design and logic BIST are other features shared by all embedded test controllers. The BIST described in this chapter facilitates sharing test resources among several memories of the same or different types.

Chapter 3, "Logic Test and Diagnosis," explains the circuit preparation required to perform an at-speed test of digital logic. Very few design rules other than the conventional scan design rules apply. The second section of this chapter introduces the hardware building blocks that are provided by the automation tools. The various test configurations and diagnostic procedures supported by the test circuitry are described. The last section describes automated solutions for addressing potential timing issues.

Chapter 4, "Embedded Test Design Flow," describes the three phases of the process of adding embedded test: test insertion in subblocks, preparation of logic blocks, top-level test insertion and integration. In each phase, the flow consists of four steps: Generate, Assemble, Analyze, and Verify. In some cases, a Prep (prepare) step is also needed.

Chapter 5, "Hierarchical Core Test," is devoted to the handling of embedded cores which are becoming prevalent in the design of large integrated circuits. The embedded test solution described in this book naturally fits the needs of a core-based design methodology. Two situations can arise. In the first situation, the core itself makes use of embedded test. A hierarchy of embedded test controllers can be implemented to test the entire integrated circuit. The second situation concerns legacy cores which are assumed to come with a predetermined test set and require access from the periphery of the integrated circuit. In both situations, great care is taken to ensure that the core and surrounding logic are completely tested.

Chapter 6, "Test and Measurement for PLLs and ADCs," introduces completely new capabilities that allow designers to accurately measure circuit parameters. The first section explains how the lock range, lock time, loop gain, and jitter of PLLs are measured using embedded test. The second section describes how to test analog-to-digital converters (ADCs) to measure their offset, gain, second and third harmonic distortion, differential non-linearity, and idle channel noise. The benefits of these solutions are explained as well as the design requirements for applying the measurement methods.

Chapter 7, "System Test and Diagnosis," presents technologies that are used for board and system tests but that are designed in the integrated circuits populating the electronic system. The first section shows how an at-speed test of board interconnect is implemented to detect and diagnose defects escaping conventional low-speed tests. The capability is compatible with existing IEEE 1149.1 standard infrastructure. Another capability built around the standard is fault insertion. This capability is extremely useful for verifying the correct implementation of the system hardware and software components dealing with online error detection, diagnosis, and recovery. A last section covers the test and diagnosis of off-the-shelf memories and their interconnections.

Chapter 8, "System Reuse of Embedded Test," describes how to use embedded test technology in a board/system manufacturing environment, or even in the field, to provide a completely self-testing, self-diagnosing system. Several architectures are reviewed and commentary on the implications for system design is included. Examples are provided of embedded test access using third-party tools.

A **Glossary** at the end of this book defines terms and acronyms pertaining to embedded test technology.

Chapter 1
Technology Overview

With recent advances in semiconductor process technology, companies that design and manufacture leading-edge products are quickly moving toward very deep submicron (VDSM) integrated circuit (IC) technology. This transition occurs because VDSM technology enables increased functionality, higher chip performance, and decreased costs.

One obstacle to achieving the full benefits of a VDSM technology is the inability of today's design and test methodologies to keep up with continual advances in semiconductor process development.

This chapter summarizes the advantages of embedded test and introduces the basic concepts underlying the embedded test solution.

Embedded Test

Conventional IC test methodologies cannot adequately and cost-effectively test the high clock speeds and millions of gates inherent in VDSM technology. The test problem is compounded by the following factors:

- Use of VDSM to create system-on-a-chip (SOC) devices

  With SOCs, the vast availability of gates is used to integrate analog circuitry, blocks of logic, memories, and processors in a hierarchical manner on a single chip.

- IC-level problems that extend to the board and system levels

  At the board and system levels, high-speed signals and high-density IC packaging challenge conventional board and system test methodologies.

- Need for shorter time to market

  As test program development time increases, the need for shorter time to market remains constant.

- Need to control test cost

  Cost-per-pin of automatic test equipment (ATE) is rapidly increasing because today's complex application-specific integrated circuit (ASIC) and SOC devices require greater performance, test accuracy, and flexibility.

As more companies experience the testing challenges imposed by VDSM technology, there is growing acceptance that semiconductor designers and manufacturers must use a hierarchical approach to testing. Leading manufacturers of high-performance products in the computing, networking, communication, and semiconductor segments of the electronics market are adopting a new hierarchical test solution—embedded test.

Embedded test extends conventional test methods with a test solution that hierarchically embeds critical portions of the ATE into the product itself. This embedded test complements the functionality of external ATE and addresses the complexity of products based on VDSM technology. The LogicVision test solution is complete and compatible with existing design and test methodologies.

Industry Trends and Conventional Test Methodologies

The increasing complexity of designing and manufacturing products, based on densely populated boards that use VDSM ICs, challenges today's conventional test methodologies.

The effects of VDSM technology on the test problem for a typical product require careful consideration. For example, a sample board contains the following design elements:

- Five 500K-gate ASICs
- At least one ASIC that is an SOC device
- Several standard IC parts
- A microprocessor
- 128 megabytes of memory
- A clock rate of 300 MHz for running the design elements
- A 66-MHz bus interface

Chips on this sample board use advanced packaging technologies, such as ball-grid arrays (BGAs). These chips are mounted on both sides of the board. Lastly, the board comprises numerous high-speed signals. Now imagine testing this complex board and its ICs using conventional test approaches such as scan/automatic test pattern generation (ATPG) and functional test.

Among the key issues to resolve, given the complexity of this board, are how to achieve a high-quality test and how to develop such a test in the shortest possible time. Because high-speed clocks are commonly employed in today's designs, at-speed test is also a requirement to address potential performance-related defects.

Using functional tests to achieve at-speed test is difficult in light of a growing bandwidth gap. This gap is the result of the limited number of external I/O and the limited external I/O frequency compared to a high number of internal functions and high internal clock rates. The National Technology Roadmap for Semiconductors predicts that with each subsequent VDSM technology evolution, this bandwidth gap will limit the scalability of tests based on external ATE. Such limited test scalability can double—and in some cases triple—the time and effort required to achieve a high-quality test using a conventional external ATE methodology such as functional test.

The same is true for board-level test development and debug. VDSM packaging and high-speed signals cause the same access and bandwidth issues, resulting in a similar impact on conventional test development time. In the semiconductor area alone, the National Technology Roadmap predicts that over one-third of IC engineering costs will be attributed to test development and debug.

Compounding the problems of VDSM test at the IC level is the emergence of SOC technology. Prior to SOC devices, companies used dedicated, function-specific testers to separately test logic, memories, or mixed-signal devices. With multiple functions now integrated onto a single device, companies must explore different solutions. They must invest in specialized external equipment such as multiple platforms with reduced throughput. Or they must upgrade to a hybrid tester that combines multi-functional testing in one ATE.

Unfortunately, these solutions are cost-prohibitive, with large multi-function IC testers costing up to five million U.S. dollars each.

What Is Embedded Test?

As Figure 1-1 illustrates, embedded test is a natural evolution of two distinct test approaches: External ATE and conventional Design for Test (DFT).

Figure 1-1 Embedded Test: a Combination of External ATE and DFT

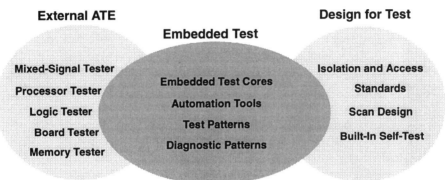

Building on conventional DFT approaches such as scan and Built-In Self-Test (BIST), embedded test integrates the high-speed and high-bandwidth portions of the external ATE directly into the ICs. This integration facilitates chip-, board-, and system-level test, diagnosis, and debug. Embedded test is implemented in two parts:

● User-configurable IP (intellectual property), in the form of design objects delivered as Register Transfer Language (RTL) soft cores

  These IP design objects implement pattern generators (either random or algorithmic), results compression, the collection of diagnostic data, and precision timing for at-speed delivery of the tests.

● A suite of powerful test automation tools running in an integrated desktop

  These tools automate generating, integrating, analyzing, and verifying test and diagnostic pattern development.

The embedded test solution cost-effectively replaces most of the ad-hoc DFT methods in use today and the time-consuming task of creating functional test patterns. Embedded test seamlessly integrates multiple disciplines:

● DFT features such as scan

● Built-in pattern generation and results collection such as BIST

● Precision, high-speed timing for at-speed test

● Test support for many different block types (logic, memories, processors, and analog circuitry)

● Capabilities for diagnosis *and* debug

Moreover, because the test technology is embedded, it provides benefits throughout the product life cycle. The aim of the embedded test approach is to design, on chip, all ATE features that are block specific or bandwidth limited by chip I/O, as Figure 1-2 on page 4 illustrates.

● On-chip data generation reduces the volume of external patterns and can be customized per block type.

● On-chip *go/no-go* and diagnostic data compression reduce ATE data logging.

● On-chip timing generation achieves true at-speed test that can scale to match process performance.

Figure 1-2 Embedded Test: Solving the Bandwidth Problem

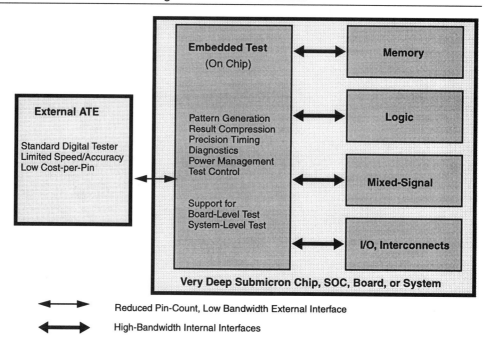

At the chip, board, and system levels, embedded test provides the following key advantages:

- Cost-effective at-speed test of logic, memory, cores, mixed-signal modules, and board-level interconnect
- Diagnostics and debug for random logic, embedded memories, mixed-signal cores, and legacy cores at the chip level
- Diagnostics and debug for I/Os, interconnects, and memory modules at the board and system levels
- Compatibility with existing flows in design and manufacturing without changes to existing infrastructures

icBIST and Embedded Test

LogicVision has developed *icBIST* —the industry's first embedded test solution for testing VDSM products. By embedding the function-specific, high-data volume and high-performance portions of both the test and the tester into the product, icBIST solves the VDSM test and diagnostic problem in a fully scalable, hierarchical manner.

LogicVision provides a comprehensive suite of design objects and EDA tools that address multiple design levels:

- At the chip level, the IP design objects and EDA tools achieve at-speed testing and diagnostics for random logic, embedded memories, mixed-signal modules, and legacy cores.
- At the board level, the IP design objects and EDA tools achieve at-speed testing and diagnostics for I/Os, interconnects, and memory modules.
- At the SOC level, the IP design objects and EDA tools support test and diagnostics of cores, including mixed-signal cores.

icBIST provides the industry's most complete suite of VDSM test and diagnostic capabilities:

● At-speed, multi-frequency, multi-clock logic BIST with cross-domain timing robustness

● IEEE 1149.1 Test Access Port (TAP) and Boundary Scan with enhanced flexibility

● Scan, testpoint insertion, ATPG, and fault simulation

● At-speed memory BIST for embedded SRAMs, ROMs, and DRAMs with unlimited ports and flexible timing

● Hierarchical isolation and access for testing cores using BIST, scan, or functional test patterns

● Mixed-signal BIST for phase-locked loops (PLLs) and analog-to-digital converter (ADC) cores

● At-speed, external memory BIST for board-level SRAM and DRAM modules

● At-speed interconnect test for board-level I/O and interconnects

● Fault insertion to verify system software diagnostics

A major strength of icBIST is the extent to which the product solves the at-speed test issue. Conventional test approaches based on scan/ATPG and functional test do little to facilitate at-speed testing of chips, boards, and systems. LogicVision's icBIST analyzes and enhances a design's at-speed test robustness, then delivers a true application-speed test—at clock speeds in excess of 100 MHz.

Figure 1-3 on page 6 illustrates a board with two complex ASICs and an array of DRAMs. icBIST fully supports all modules embedded in these two ASICs, as well as the external DRAMs, board-level I/Os, and interconnects.

Figure 1-3 LogicVision and Embedded Test

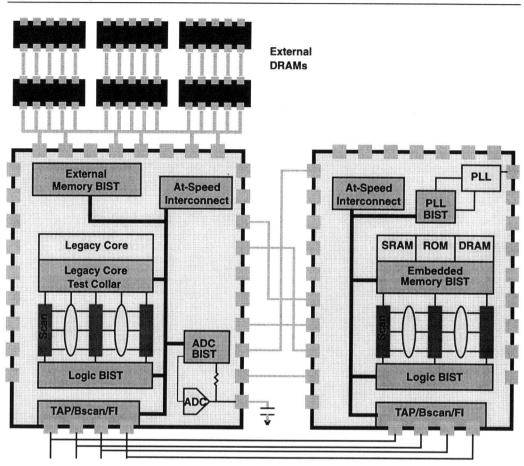

DFT Methods Used in Embedded Test

DFT methods have been used for many years to help augment the testability of chips, boards, and systems. These methods provide access and isolation, enabling test application. They also increase the effectiveness and coverage achieved by the test.

DFT solutions can be used with any type of testing: functional, scan, or embedded. Although the use of DFT has an impact on the size of designs, with the size and complexity of today's designs, this impact becomes less significant. Additionally, the time to market and cost payback of using DFT has improved tremendously.

This section provides an overview of the basic DFT concepts and technologies that embedded test and icBIST employ:

- Structural versus functional test
- BIST (algorithmic and random pattern)
- IEEE 1149.1 TAP and Boundary Scan
- Scan design for user logic

Structural Versus Functional Testing

Test processes can be divided into two categories:

- *Functional* testing—checks some aspect of the expected design behavior, for example, performance at application clock speed.
- *Structural* testing—uses a pattern set that has a known fault coverage. This testing is used for manufacturing.

Functional testing verifies the input-to-output behavior of the circuit, whereas structural testing involves strategies to verify the individual elements of the design: logic gates, transistors, and interconnects.

Some test processes achieve both functional and structural objectives. For example, many RAM test procedures primarily achieve a structural test objective. Because these test procedures do not require any modification of the circuit under test, they can additionally achieve a functional objective.

Often, it is tempting to augment functional tests that verify the design behavior in an effort to mold these tests into manufacturing (structural) tests. However, it is difficult to meet target test coverage using modified functional tests because of low controllability and observability of signals, lack of automation, and lengthy fault simulation times.

Functional tests are not reusable from one level of packaging to another. Functional tests also requiring rewriting, which increases time to market. The fault coverage from functional tests decreases significantly at the board and system levels because controllability and observability decrease, which potentially affects the resulting product quality. Moreover, because the functional test used for manufacturing requires intimate knowledge of the function, the task of deriving the test for a particular function is not easily delegated to a test engineer who did not participate in its design.

Structural tests address the shortcomings of functional tests used for manufacturing. High fault coverage is achieved by partitioning the circuit under test into subblocks. These subblocks guarantee high controllability and observability for any signal.

- When DFT methods such as scan are used, tests can be automatically generated using tools such as ATPG.
- When DFT methods such as BIST are used, tests can be automatically generated and reused at various levels of packaging using standard test access methods.

Embedded test implements a structural-based test methodology. The resulting circuit remains suitable for functional test. Embedded test uses design objects and test programs that verify up to 100 percent of the gates, transistors, and interconnects in the circuit under test.

Table 1-1 summarizes the benefits of using structural testing over functional-based tests for manufacturing.

Table 1-1 Comparison of Structural and Functional Tests

Structural Testing	Functional Testing
Verifies individual logic gates/transistors	Verifies input-to-output behavior
High fault coverage	Low-to-medium fault coverage
Largely automated process	Mostly manual process
Reusable tests	Non-reusable tests
Standard test process	Custom test process

Built-In Self-Test (BIST)

BIST is a hierarchical DFT strategy that reduces the need for external test. With BIST, a micro tester, complete with a pattern generator, is brought onto the chip, enabling the chip to test itself. Although this tester requires more silicon area, the savings realized through automation of the testing process, higher quality of testing, and reuse at all levels in the design hierarchy (core to chip to board to system) make this DFT method very attractive.

As Figure 1-4 illustrates, BIST adds logic to the circuit under test to realize pattern generation, pattern capture and compare, and self-test control. Typically, BIST circuits provide low-speed testing and indicate only a passed or failed test.

With the embedded test approach, LogicVision's BIST circuits provide high-speed operation to implement at-speed test, as well as different levels of diagnostic data. This implementation is described more fully in the section "Capabilities of icBIST" on page 17.

Figure 1-4 Logic Added for Self-Test

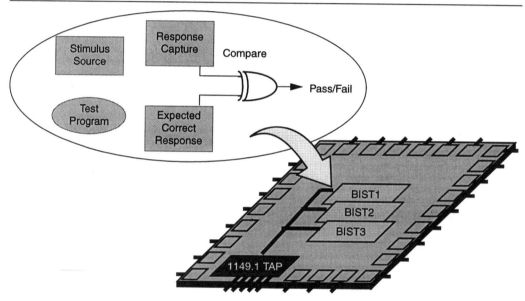

The appropriate type of controller is used with each type of embedded core—for example, memory, logic, socket, or phase-locked loop. BIST uses controllers that are typically accessed via a direct, low-pin-count and low-speed interface or through the five-pin IEEE-standard 1149.1 TAP. Access using the TAP is recommended because it simplifies BIST reuse at the board and system levels. In both cases, the protocol of the controller is the same:

1. Initialize BIST operations.

2. Monitor status.

3. Read results.

The key benefits of BIST are faster time to market, higher-quality testing, and lower costs, as explained below.

Faster Time to Market

BIST automates a high degree of the test development process and simplifies the development of test programs.

BIST fits right into the design flow at the RTL level (Verilog or VHDL). Test logic is optimized at the same time as user logic—during the synthesis process—eliminating iterations between the gate and RTL levels.

BIST supports an embedded core environment. Cores that have BIST are inherently self-testing, and the test program is self-contained, saving time with each reuse.

BIST tests are easy to reuse at all levels of packaging (wafer, packaged chip, board, and system) because of standard access methods. Prototypes are easier to debug, enabling manufacturing to start earlier.

Higher-Quality Testing

BIST algorithms are targeted specifically at known defect types, resulting in high fault coverage. For example, memories and mixed-signal functions require specific fault models. Escape rates at wafer probe are reduced with corresponding savings on packaging costs (associated with packaging bad die).

Chips can be tested at speed without incurring yield losses because of tester inaccuracy. These losses are increasing with each generation of the VDSM process because the clock period of chips is approaching tester accuracy limits.

Because a combination of BIST and *boundary scan* (see the "Test Access Port and Boundary Scan" section on page 10) allows the designer to test chips by connecting only a few signal pins to the tester, yield losses caused by poor contact at the wafer probe are significantly reduced. This is becoming an important factor because the number of chip pins can be in the thousands.

Chips wear out. Infant mortality, electro-migration, circuit oxide breakdown, hot electron effect, and die metallization corrosion are all potential causes of failure in the field. BIST can be used to determine bad parts during system manufacture and to diagnose problems in the field.

Some problems, such as temperature and noise (signals and supply) problems, can be reproduced only when the chip is operating in the system. Such design margins are difficult conditions to reproduce during manufacturing. BIST can be used to pinpoint chips that are affected by environmental conditions and therefore reduce the occurrences of *No Fault Found* (NFF).

Lower Costs

BIST reduces the dependency on expensive ATE. Today, high-end ATE costs are approximately five million U.S. dollars. This cost is attributed to the need to add more, and faster, pin electronics. It is also attributed to the need for more memory to store the larger patterns required for complex chips. BIST solves the ATE cost problem by moving data—formerly loaded through pins by the ATE—*directly onto the chip.*

The need for multiple, external ATE lines is also diminished by using a single, external ATE line and multiple, function-specific, on-chip BIST circuits.

Test Access Port and Boundary Scan

Another DFT technology used in embedded test is the IEEE 1149.1 Test Access Port (TAP) and Boundary Scan. This technology, developed in the late 1980s by the Joint Test Action Group (JTAG), was standardized in 1990 by the IEEE [1, 2]. It defines a chip-level test structure that is used to implement a simplified board and system-level test strategy.

Since the mid-1970s, structural testing of printed circuit boards (PCBs) has relied heavily on the use of the in-circuit, *bed-of-nails* test technique. This testing method uses a fixture that contains a bed of nails for accessing individual devices on the board. Contact is made through test lands laid into the copper interconnect or other convenient contact points. Fundamentally, the in-circuit, bed-of-nails technique relies on physical access to devices on the board. For plated-through-hole technology, access is usually gained by adding test lands to the interconnects on the solder side of the board.

The advent of surface-mount devices enabled manufacturers to place components on both sides of the board. Smaller pitch between the leads of the surface-mount components caused a corresponding decrease in the physical distance between interconnects. This decrease impacted the ability to place a nail accurately onto a target test land. The development of multi-layer boards further compounded access.

The JTAG effort was motivated by two goals:

- Create a standard mechanism for testing ICs and the interconnect between them.
- Reduce the total test cost of a system.

The IEEE 1149.1 standard is based on the concept of a serial shift register around the boundary of the device. The term *boundary scan* derives from this concept. Adherence to the IEEE 1149.1 standard facilitates interoperability between chips from different vendors. The standard is supported by numerous suppliers of commercial tools and ATE equipment.

Boundary-scan (BScan) design is a DFT method that is applied to the input and output pin circuitry of a chip to enhance accessibility and testability. Through BScan, each I/O bonding pad is associated with a combination of latches, flip-flops, and multiplexers that, in test mode, can be configured in several ways to perform the following basic test-mode functions:

- Sample or force data on inputs.
- Sample or force data on outputs.
- Force input data and sample output data on bidirectionals.
- Selectively disable or enable each output.

The IEEE 1149.1 standard provides for boundary-scan cells and a five-pin TAP. These pins include the following:

- *TDI*—Test data-in
- *TDO*—Test data-out
- *TCK*—Test clock
- *TMS*—Test-mode select
- *TRST*—Test reset (optional)

These pins provide a standard protocol for loading and unloading TAP instructions, boundary-scan chains, and internal scan chains, as Figure 1-5 on page 11 illustrates. The protocol also defines standard TAP instructions for test operation modes.

Figure 1-5 BScan Access Using the Five-Pin Tap Interface

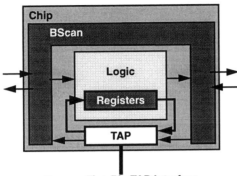

Four or Five-Pin TAP Interface

Essentially, boundary scan adds scannable logic cells to the pins of a chip. These BScan cells provide two test modes:

- *Internal test mode*—enables control and observation of the *internal functions* of the chip via the boundary-scan cells. The internal circuitry is tested while the chip is isolated from external pins.

- *External test mode*—enables control and observation of the *pins of the chip* via the boundary-scan cells. Off-chip testing and DC tests are provided to verify chip attachment and interconnections.

BScan simplifies both interconnect and chip test at the board level by partitioning. In the same way that scan design enables automatic test generation for internal chip logic, BScan enables automatic test generation and diagnostics for the board interconnect network. In some cases, BScan can be used to test the *glue logic* between chips. (Glue logic is miscellaneous, random logic that is not application related but is necessary for connecting elements of a design.)

BScan is more reliable and easier to use than a bed-of-nails test because it is a contactless method. DC parameters (and some AC parameters) can be measured automatically at the chip level using the BScan chain. The design cost of using BScan is a larger and potentially slightly slower input/output (I/O) peripheral cell, depending on its design.

BScan cells are typically provided for bidirectional, input, output, and enable applications.

- BScan cells for bidirectional pads and input pads are the most commonly used.

- Output pads are often replaced with bidirectional pads when using BScan because outputs do not allow BScan to capture signals at the pad.

- The BScan enable cell is used to control functional-enable signals, such as the enable of the bidirectional pads.

- Special, sample-only BScan input cells are also used in applications whose signals cannot have their functionality changed by a BScan cell. An example of such an application is clock nets.

Figure 1-6 on page 12 illustrates a generic BScan cell. This logic block is placed between the pad (and associated protection devices, level shifters, and drivers) and the internal chip circuitry.

- For an *output* pad, DIN (data input to BScan cells) comes from the internal chip circuitry and DOUT (data output from BScan cells) goes to the pad.

- For an *input* pad, DIN comes from the pad and DOUT goes to the internal chip circuitry.

Figure 1-6 Generic Boundary-Scan Design

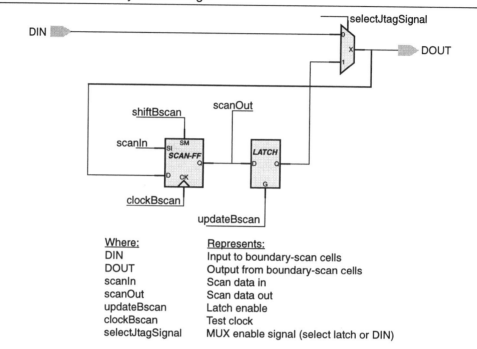

Where: Represents:
DIN   Input to boundary-scan cells
DOUT   Output from boundary-scan cells
scanIn   Scan data in
scanOut   Scan data out
updateBscan   Latch enable
clockBscan   Test clock
selectJtagSignal   MUX enable signal (select latch or DIN)

Function and symbol definitions used in Figure 1-6 are as follows:

- *SCAN-FF* is a dual, edge-triggered scan flip-flop. The output changes on the rising edge of *clockBscan* and the input is captured on the falling edge of *clockBscan* for robust timing. The polarity of *clockBscan* is adjusted to the test mode (external or internal) in effect.

- *LATCH* refers to non-inverting latches that are transparent when the latch enable is high. When *shiftBscan* is high, the BScan cells are configured as a shift register with *scanIn* coming from the previous stage and *scanOut* going to the next stage. Therefore, the purpose of the latch is to prevent *DOUT* from rippling when data is shifted through the register, so *updateBscan* is typically low during shift mode.

- The multiplexer controlled by the signal *selectJtagSignal* selects between normal data for output (*DIN*) or the BScan data. The signal is called *selectJtagInput* for an input pad and *selectJtagOutput* for an output pad.

- The multiplexer controlled by the signal *shiftBscan* in *SCAN-FF* selects shift register mode or sample mode. The normal configuration is *sample mode*, in which the flip-flop captures the value at DIN.

The implementation described in Figure 1-6 is typically available in icBIST. Other implementations are possible to accommodate special requirements.

All the boundary-scan cells are configured into a parallel-in, parallel-out *shift register.*

- A *parallel-load* operation—referred to as a *capture* operation—loads into BScan input cells those signal values arriving at device input pins. The operation also loads into BScan output cells those signal values passing from core logic to device output pins.

- A *parallel-unload* operation—referred to as an *update* operation—passes signal values already present in the BScan input cells into the core logic instead of passing the values on chip pins. This operation also passes signal values already present in the BScan output cells to chip output pins instead of passing the output values generated by the core logic.

Data can also be shifted through the shift register in serial mode, starting from a dedicated *test data-in* (TDI) chip input pin and terminating at a *test data-out* (TDO) chip output pin. A *test clock* (TCK) is fed through another dedicated chip input pin. The mode of operation is controlled by a dedicated *test mode select* (TMS) control input pin.

A BScan cell needs several control signals, aside from the clock. These signals generally come from a TAP. The TAP serves as the interface for the TDI, TDO, TCK, and TMS signals and generates any additional internal signals needed to control the boundary scan and other test modes on the chip.

By applying the appropriate pattern to the TDI, TDO, and TMS pins, the TAP can be used to place the chip into any desired test mode. The standard specifies a number of default test modes, including internal test mode, external test mode, and bypass mode.

The key benefits of the IEEE 1149.1 TAP and boundary-scan combination are the following:

- Provides common access protocol for all test features in the chip (for example, BIST controllers and user-defined test registers for debug).

- Supports manufacturing test at multiple levels through the internal and external test modes.

- Enables low-cost chip test and burn-in, requiring contact of power pins, TAP interface pins, and clock pins. Contact occurs through low-cost testers and probes. Such testing applies to all levels of packaging of the product (wafer, chip, board, and system).

- Reduces test pattern development and debugging time at all levels of packaging, resulting in faster time to market. Pattern development and debugging are reduced because access to all chip test features is identical for all chips. Additionally, the IEEE 1149.1 standard is universally supported by many tool and ATE vendors.

Scan Design for User Logic

Scan design is a fundamental DFT method involved in embedded test that enables structural test. Scan design can be used for automating external scan/ATPG testing and logic BIST by providing a high degree of controllability and observability, high fault coverage and diagnostics, and independence from functional complexity.

Two factors are critical to implementing scan design successfully: the clocking strategy used in a system and the scan design style. Gaining familiarity with the following scan design guidelines and scan rules before implementing logic BIST will help designers obtain the best results in the shortest amount of time.

A *scannable circuit* is a circuit that provides a mode in which all storage elements in the circuit are chained together into one or more shift registers, with external input and output. Using this single shift register, the following functions occur:

- Shift an arbitrary state into the memory elements.

- Operate the circuit normally for one or more clock cycles.

- Shift out the resulting state of the memory elements.

Using scan design enables high fault coverage because of the high degree of controllability and observability that this technology provides.

In a scannable circuit, memory elements and combinational logic driven by the memory elements can be treated separately for testing. The storage elements can be tested by shifting a simple pattern through the register and then observing the outputs.

With scan design, external test patterns for combinational logic can be generated automatically using ATPG tools in a very efficient manner. Once generated, the combinational logic tests can be applied by shifting inputs and outputs via the scan chain.

Scan design also enables other DFT methods such as BIST, because it provides a simple mechanism for applying and collecting test patterns and results. Another advantage is that a scan circuit is inherently easy to diagnose. This feature is invaluable when a root-cause analysis of test failures is required, whether for external test or BIST.

Scan Styles

Although there are different scan design styles, *multiplexed scan* is the most popular style and the only one discussed in detail in this book. Multiplexed scan simplifies at-speed testing because the test clock and the functional clocks use the same clock distribution network.

To implement multiplexed scan, it is necessary to add an extra multiplexer on the input data path of each flip-flop, as Figure 1-7 illustrates. This multiplexer is incorporated within the scannable flip-flops available in the cell libraries. These scannable flip-flops are usually slightly larger and the setup time is typically increased slightly.

Figure 1-7 Scannable Flip-Flops

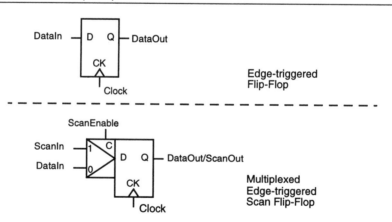

All such scannable flip-flops must then be connected together to form one or more scan chains. icBIST automatically substitutes non-scan flip-flops with scannable flip-flops and connects these flip-flops into scan chains.

Figure 1-8 on page 15 shows a small circuit before scan is inserted.

Figure 1-8 Portion of a Circuit Before Scan Insertion

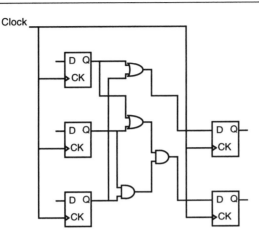

Figure 1-9 shows the same circuit after scan is inserted.

Figure 1-9 Portion of a Circuit After Scan Insertion

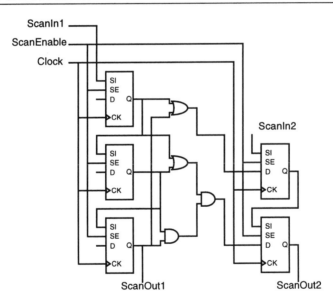

Scan Protocol

Figure 1-10 on page 16 illustrates a typical scan sequence. The clock is free-running and the *ScanEnable* signal is kept high for the number of clock cycles that is sufficient to load new values into all flip-flops. Three clock cycles are sufficient for the circuit in Figure 1-8. Normally, hundreds—or even thousands— of clock cycles are necessary, depending on the length of the longest scan chain.

The average chain length can be determined by dividing the total number of flip-flops by the number of scan chains. External ATE usually supports a limited number of scan chains, which can increase the average chain length and result in longer test times and larger scan buffer memory.

On every clock cycle, the output of each flip-flop can potentially change (50-percent probability). Additionally, the output of the combinational portion of the circuit can also change to respond to the new input. However, during the scan shift mode, the output of the combinational circuit is not captured. It is only captured one clock cycle after the last shift operation of the scan chain.

- The clock cycle during which the scan-enable signal is low is called the *capture cycle*.
- The clock cycles during which the scan-enable signal is high are called *shift cycles*.

Figure 1-10 Sequence of Operations

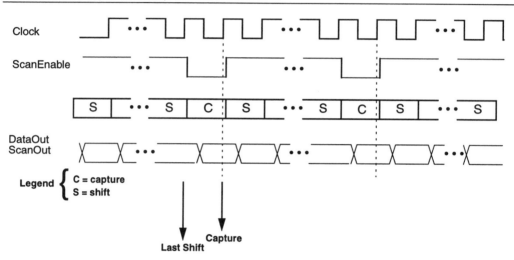

Scan Design Restrictions

The following design features require special consideration during the design phase. Minimizing the use of such features can achieve higher fault coverage and simplify test development. Note that in order to implement BIST, scan design rules are slightly more restrictive.

- **Gated clocks**. Clocks should be directly controllable from primary input sources. Static multiplexing or other forms of gating are allowed to achieve this objective in test mode. Clocks driven from inputs can be dynamically gated with signals derived from flip-flop outputs. However, clocks cannot be generated in the core logic during test mode (for example, ripple counters).

- **Asynchronous sets and resets**. These signals must be disabled during BIST. Typically, only one flip-flop should be asynchronously reset per clock domain. All other flip-flops can be reset synchronously from the asynchronously reset flip-flop. The fault coverage loss is minimal and ATPG patterns, not reusable in the system, can be used to test the asynchronous sets and resets that are used.

- **Internal tri-state bus drivers**. Tri-state buses must be replaced with multiplexers whenever possible. If tri-state buses cannot be avoided, the enable signals of bus drivers must be uniquely decoded. Optionally, a disabling mechanism can be used during shift mode and pull-ups can be connected to these buses for higher fault coverage.

- **Set/reset latches**. Set/reset latches must be replaced with flip-flops.

- **Transparent latches**. Transparent latches must be used for retiming only. All other transparent latches must be replaced with flip-flops or made transparent during test mode.

Clocking Issues

Another important consideration of scan design is how clocks are distributed and connected to flip-flops.

For each clock domain, all flip-flops must be connected to the same level of the clock distribution network to minimize clock skew. The flip-flops within the clock domain can then safely transfer data between themselves in both the scan shift and scan capture modes of operation.

All flip-flops must update their output from the same clock edge within a clock domain. However, it is legitimate to use flip-flops clocked on the opposite edge to interface with flip-flops in a different clock domain. It is also acceptable to use transparent latches for retiming between clock domains.

In many applications, it is common to have clock sources that are completely asynchronous to each other.

- During the normal operation of the circuit, asynchronous interfaces are designed to transfer data between clock domains to ensure that clock skew is not an issue.

- During scan, the entire circuit is operated in a way that all clock sources are driven from a single test clock, or derivatives of that clock, and the functional asynchronous interfaces are disabled. In this case, clock skew can make transferring data between flip-flops of different clock domains unreliable by causing hold-time violations.

LogicVision's icBIST implements a patented approach to solving this common scan problem. This icBIST feature, referred to as *capture-by-domain*, identifies all data transfer paths that are potentially unreliable and proposes circuit modifications to make the data transfer completely robust.With the designer's approval, the capture-by-domain feature can also implement the circuit changes.

The behavior of flip-flops connected at both ends of unreliable paths is modified so that these flip-flops never capture at the same time (that is, during the same capture cycle). In this way, no hold-time violations can occur. This modification does not affect test time or functional timing. The cost is limited to one extra multiplexer for each flip-flop at the source of the paths of interest.

The capture-by-domain method of fixing hold-time violations is preferable to using flip-flops clocked on opposite edges. The capture-by-domain method works at higher frequencies, which is an important consideration for at-speed logic testing.

There is no strict limit on the number of clock sources. However, it is best to minimize the number of clock sources to reduce the number of interfaces between clock domains.

Scan Design Cost

The relative cost of adding scan to a design is directly proportional to the number of flip-flops in the user logic, because each of these flip-flops must be replaced by a slightly larger scannable equivalent. This cost can vary between 5 to 10 percent of the total logic's gate count without scan.

Capabilities of icBIST

This section describes the complete set of embedded test capabilities supported as options in LogicVision's icBIST product. icBIST provides a broad set of capabilities that implement embedded test for chips, systems-on-a-chip, boards, and systems. Each capability provides user-configurable, function-specific IP in the form of design objects and automation tools to perform these functions:

- Generate and insert the design objects into the chip
- Check design rules
- Create testbenches to verify the embedded test modes
- Support standard Verilog and VHDL design flows

The design objects provided by icBIST are generated by automation tools in order to create an optimized design implementing only those features needed for the targeted chip. This flexibility ensures a minimum gate overhead with the exact set of required features.

All icBIST capabilities are embedded at the chip level. Some capabilities focus on the test and diagnosis of on-chip features. Other capabilities are off chip, focusing on test and diagnosis at the board and system levels. All icBIST features are implemented as synthesizable digital designs, including the mixed-signal features for PLL and ADC.

icBIST Architectural Framework

Figure 1-11 illustrates the complete icBIST chip-level design hierarchy. Some of the test controllers are instantiated at the top level of the circuit, whereas the other controllers and collars become part of the user core block. There is no restriction on the core block hierarchy, and more than one user core block can be present at the top level.

The collared cores of Figure 1-11 can themselves contain BIST controllers or scan chains that are accessible in a hierarchical manner. The arrangement of the test features within a core is similar to the one of the chip top level. Legacy cores that do not include any structural test capability are equipped with a collar that makes possible the application of functional vectors from chip pins. All collars provide the ability to test the interface between the cores and the surrounding logic.

Figure 1-11 Chip-Level Design Hierarchy

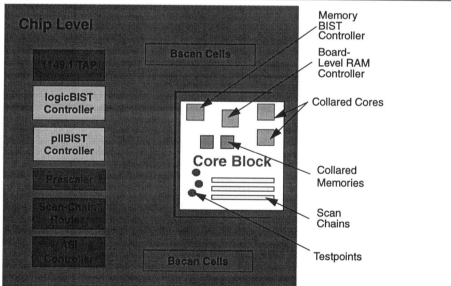

At-Speed Logic Test

icBIST implements an at-speed test of logic using a patented, at-speed, multi-frequency, multi-domain logic BIST approach. The logic BIST capability is based on the extension of a scan circuit (usually full scan) into a self-test version. This self-test is performed using a pseudo-random pattern generator (PRPG) as stimuli generator and a multiple-input signature register (MISR)-based cyclic-redundancy checker (CRC) for output results compression. The logicBIST controller consists of the PRPG, the MISR, and a state machine capable of operating at full application speeds.

In addition to the RTL design objects for the logic BIST controller, the logic BIST capability also provides RTL design objects for the IEEE 1149.1 TAP and boundary-scan cells, a scan chain interface (called a chain router), a clock prescaler, and testpoints, to implement a more random pattern-sensitive scan design. If logic BIST is used on a core or a subset of a large chip, the TAP and boundary-scan cells are replaced by collar elements that provide the same level of isolation and access.

The logic BIST capability supports three different modes of logic testing: at-speed logic BIST, multiple external scan chains, and scan through TAP. The logic BIST mode is normally used as the default, and the scan modes are used as a backup or supplemental strategy.

The logic BIST automation tools support generation and integration of the IP, rules checking and analysis, scan chain and testpoint insertion, verification, and test generation for all three modes. The logic BIST capability supports LogicVision-supplied combinational ATPG and fault simulation or the use of a third-party ATPG for scan-pattern generation.

logic BIST Operation

The sequence of operation for logic BIST is the following:

1. Random patterns are applied to the circuit through scan data-in signals to the scan path.

2. Responses are taken into the MISR from the scan data-out outputs from the scan path.

3. At the end of the self-test cycle, when the last random pattern has been applied, the final signature is read from the MISR and compared with the expected final result to produce a pass/fail decision. Optionally, the signature can be scanned out for comparison to a reference signature by the external ATE.

Figure 1-12 on page 20 shows the application of logic BIST to a circuit with multiple scan chains, which include boundary-scan chain segments and internal scan chains. The scan chains can be clocked at different frequencies to allow at-speed testing of the circuit. The logicBIST controller derives subfrequencies (*Clock[2]* and *Clock[3]*) from a system clock.

icBIST supports multi-cycle paths, a design technique often used in high-speed circuits. The logicBIST controller generates clock-enable signals (*ClockEnable[2]* and *ClockEnable[3]*) that correspond to the number of clock cycles required to propagate signals between flip-flops in a multi-cycle path. Circuit synchronization provides robust timing at high speeds and along signal paths that cross different clock domains.

A boundary-scan chain that comprises scannable flip-flops located near the primary input and output pins is used to isolate the circuit so that the output response does not depend on the state of the primary inputs. This condition is important at chip, board, and system levels.

●   At the chip level, the tester does not have to store test data for all primary inputs. The tester simplifies the synchronization of the test because there is no source of input stimulus other than the BIST controller.

● At the board and system levels, the inputs are not easily controllable. If these inputs are controllable, it is preferable *not* to rely on them for diagnostic purposes. The boundary-scan chain prevents unknown values from affecting the output response in order to make the test repeatable.

During logic BIST, all output drivers are disabled by default to minimize power consumption during BIST and to avoid bus contentions or other hazards at the board or system level.

Figure 1-12 Logic BIST Approach

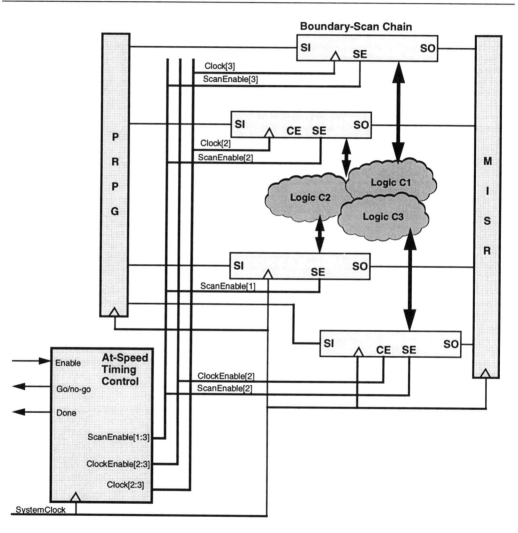

logic BIST Clocking

A main feature of logic BIST is its capacity to test a complete chip with clock domains running at different frequencies, testing at speed all clock domains and the interfaces between each of these domains. This feature also reduces test time because all portions of the chip are tested simultaneously. This feature is patented and is exclusive to LogicVision's icBIST product.

Typically, the clock selected as the main test clock operates at the highest frequency in the system. The logicBIST controller is connected to the corresponding clock tree. This controller generates two subfrequencies from the main test clock—*Clock[2]* and *Clock[3]*—that run at one-half (1/2) and one-quarter (1/4) the rate of the high-speed frequency, respectively. Figure 1-13 on page 21 shows the timing of the signals from the logicBIST controller.

Figure 1-13 Sequence of Operations (Multiple Frequencies)

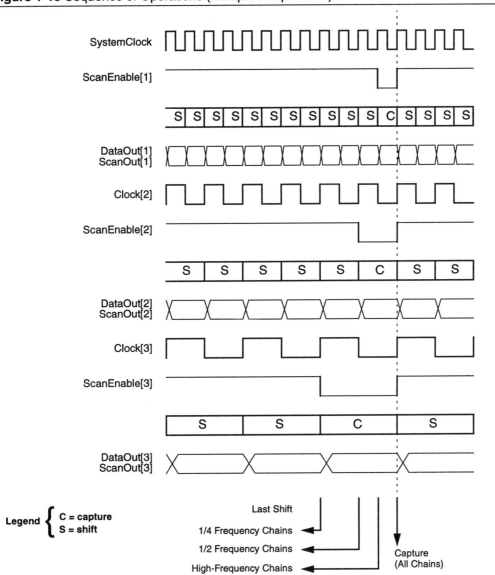

For each frequency, the corresponding clock and scan-enable inputs and flip-flop outputs are depicted. A symbolic representation of the shift and capture cycles is also included. The three groups of clock domains (several domains can run at the same frequency) are synchronized so that all flip-flops enter the capture cycle of the scan sequence at the same time, within clock skew limits.

The last shift cycle occurs at different times depending on the frequency for a particular scan chain. The clock frequency of the source flip-flop determines the time allowed to propagate a signal from a flip-flop output to any flip-flop input. The time that occurs between the clock edge causes the last shift of the source flip-flop, and the clock edge causes the capture at the destination flip-flop.

The section "Clocking Issues" on page 17 described how to neutralize the effect of clock skew between clock domains by using the capture-by-domain feature. If the skew between two clock domains is very large and becomes comparable to or greater than the clock periods of the two domains, the effective frequency of the source flip-flops must be reduced. To address this issue, icBIST supports the use of multi-cycle paths.

Figure 1-14 shows a simple incrementer that uses multi-cycle paths for fast operation. All paths that originate from output *Q[0]* need to settle within one clock period. However, all paths originating from outputs *Q[1]* to *Q[3]* are less critical because they have at least two clock cycles to settle from the time they change.

Figure 1-14 Multi-Cycle Path Example

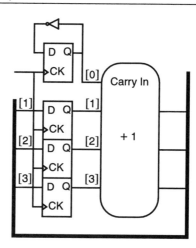

Reducing the number of critical paths helps the synthesis tools reduce the area of the circuit and optimize its performance.

Multi-cycle paths have always been perceived as a fundamental problem for BIST because most schemes cannot handle them. Signatures obtained from the circuit can vary depending on the test frequency. Frequency-dependent variation is not an acceptable outcome. LogicVision provides a solution in which, after the source of multi-cycle paths is identified, icBIST automatically inserts special flip-flops that extend the duration of the capture cycle to match the length of those paths. The flip-flops that source multi-cycle paths are controlled by the appropriate *ClockEnable* signals (refer to Figure 1-12 on page 20). This scheme allows the circuit to be tested at its rated speed and with signatures that are always consistent and not frequency dependent.

As mentioned earlier, in the section "logic BIST Clocking" on page 20, transferring data signals between clock domains when the clock skew is comparable to the clock periods of the two clock domain sources can be handled in the same way as that described for multi-cycle paths. In this case, the strategy for handling multi-cycle paths is to use the capture-by-domain feature.

Fault Coverage and Quality

A large number of random patterns applied at high speed results in higher quality for logic if the stuck-at fault coverage is high. This is because each node can be controlled to a logic zero value or logic one value using several combinations of different input patterns. These patterns potentially sensitize faults not modeled by the stuck-at fault model. The same node is also easily observable to propagate the fault effect to an output. The more often a node is observed, the more chances there are to detect unmodeled defects or what are called *collateral faults*.

icBIST also measures the transition fault coverage to provide an objective metric of the coverage of delay defects and stuck-open defects in CMOS circuits. Slow-to-rise and slow-to-fall faults are associated to each input and output of all gates. To detect a slow-to-rise fault at a gate output, it must first be initialized to a logic zero value with the first pattern and to a logic one value with the next pattern. The transition must be propagated to an output (or flip-flop input) to be detected.

In a scan-based test, the initialization pattern is applied on the next-to-last shift. The pattern that causes the transition is applied on the last shift cycle. Because the test is applied at speed, the transition has exactly one system clock cycle (or several cycles if it is a multi-cycle path) before it is observed, therefore detecting potential delay faults.

Diagnosis

With icBIST, several levels of diagnosis are available. Signatures, seeds and all other registers of the BIST controller can be loaded and inspected through a serial interface for diagnostic purposes. Using scan vectors, the BIST controller itself can be tested along with the entire circuit.

Two features are used to diagnose speed-related failures.

- The clock prescaler can reduce the system clock frequency before it is applied to the BIST controller and the circuit under test. When the circuit with the BIST fails in the system, this feature is useful for determining if the failure is speed related.

- A capability called *clock freeze* can stop the system clock after a capture cycle to shift out the contents of all flip-flops of the circuit and log them onto a tester at low speed. Using this feature, it is possible to diagnose faults that manifest themselves only at high speed. Because the contents of all flip-flops are calculated by icBIST, the failure can be isolated to one or more flip-flops whose capture data is incorrect.

Trade-Offs

Several factors determine the trade-off between fault coverage and test time:

- Test clock frequency
- Number and length of scan chains
- Number of patterns
- Maximum power levels allowed at different packaging levels (wafer, chip, board, or system)

Although scan logic is significantly more sensitive to random patterns, it is possible that the target fault coverage cannot be achieved within the test time allowed. In this case, the maximum number of patterns is greatly exceeded; hence, simply adding scan chains is not practical. To solve this problem, icBIST uses testpoints to increase the controllability and observability of the circuit and reduce the number of test patterns.

Figure 1-15 shows the characteristic fault coverage curve for a set of randomly generated patterns applied to a circuit before and after testpoint insertion where the targeted number of patterns is not enough to realize the desired target coverage. After an initial surge of coverage with the early patterns, the curves level off and eventually become essentially flat. This behavior is typical of random pattern testing methods. Testpoint insertion is one of the techniques that gives the designer control over test time by increasing the rate of fault detection in the quasi-flat region.

Figure 1-15 Effectiveness of Testpoint Insertion

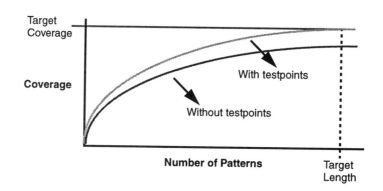

Power

By default, all output drivers are turned off to minimize power consumption during logic BIST. However, the power consumption of the chip can still exceed the maximum ratings when testing at speed. This is due to the higher level of activity of the circuit caused by the application of random patterns and the insertion of testpoints.

logic BIST has an exclusive run-time programmable power-level feature that allows control of the average power without sacrificing at-speed testing. This is done by shifting most of the scan patterns at a programmable low speed and applying the last few shift cycles and the capture cycle at the maximum speed.

Benefits

In addition to the general benefits listed in the section defining BIST and applicable to all LogicVision BIST technologies, the following benefits deserve specific mention for the logic BIST capability.

- *High quality of final product*—logic BIST achieves very high fault coverage because it is based on scan, uses random patterns and testpoints, and is applied at speed.

- *Short test pattern development and debugging time*—logic BIST automation includes the generation and insertion in the design of the logicBIST controller, scan chain and test point insertion, and identification and correction of unreliable timing paths.

- *Easy debug-and-diagnostics at the chip level*—logic BIST is scan based, and provides full control and observation of the controller's internal registers (pattern length, seed of pattern generator, signature register, and so on), full observation of internal scan chains (clock freeze feature), and full control of the clock sources (system clock pin, PLL output, and TAP clock).

- *Short test times for decreased manufacturing costs*—logic BIST can test the entire chip at speed. It supports the use of a very large number of scan chains and the insertion of testpoints to achieve coverage with fewer patterns.

Logic BIST can achieve higher quality with fewer tester resources than with stored scan patterns. For example, consider a circuit with 1 million gates and 50,000 flip-flops that runs at 100 MHz. Table 1-2 shows that a significantly higher fault coverage is achieved in half the time it takes to apply scan/ATPG patterns by increasing the number of chains and test frequency and possibly by adding testpoints.

Table 1-2 Results of Logic BIST

	SAF Coverage	Transition Coverage	Unmod. Coverage	Number of Patterns	Tester Storage (bits)	Number of Chains	Test Time (s)
Scan/ ATPG	100%	80%	low	2000	200M	8	1.25 @ 10 MHz
logic BIST	100%	90%	high	64000	2K	50	0.64 @ 100 MHz

Embedded Memory Test

icBIST also includes the memBIST-IC option for at-speed test of embedded memories. Increasingly, designers are using more embedded memories within their designs. Typically, additional circuitry is required to allow the external tester access to each embedded memory through the external I/O pins of the IC. It is becoming more difficult to provide access without impacting chip routability, area, performance, and design cycle time.

The memBIST-IC option eliminates the design effort needed to make embedded memories accessible for external tests and provides at-speed self-test using standard memory testing patterns. memBIST-IC provides a patented, deterministic BIST for embedded single and multiple-port SRAMs, ROMs, and DRAMs. The serial data interface simplifies communication between multiple embedded RAMs and a BIST controller. For testing very large memories or to achieve very short test times, memBIST-IC also supports a more conventional parallel data interface.

Memory BIST Operation

memBIST-IC generates the following:

- Customized memory BIST controller (RTL design object)
- Individual collars for each RAM (RTL design objects)
- Simulation testbench (VHDL or Verilog) for verification
- Synopsys shell scripts for logic synthesis of the RTL

The insertion of the collared memories and the BIST controller into the design is completely automated. Each memory—located anywhere in the design hierarchy—is collared and connected to the controller.

Figure 1-16 on page 26 illustrates LogicVision Memory BIST.

Figure 1-16 Memory BIST Concept

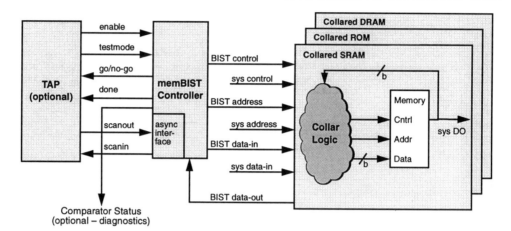

Comparator Status
(optional – diagnostics)

Only one memory BIST controller is needed per design, regardless of the number of memories, type, size, and number of ports. For high-speed testing, LogicVision recommends using one controller per clock domain. One controller per clock domain eliminates timing problems between the controller and the memories because of clock skew. For multi-port memories connected between two clock domains, the controller should receive its clock from the clock domain with the highest frequency.

memBIST-IC includes a feature that optimizes sharing of the BIST controller. Memories that have a different number of words can be tested in parallel in a single step, thereby reducing test time. This feature is especially important when testing the data retention capability of the memory.

Algorithms and Coverage

The memory BIST controller implements an order 22N March test, which is serialized to dramatically reduce routing and gate area yet allow for at-speed BIST operation. The algorithm covers a large number of defects (stuck cells, address decoder faults, all two-bit coupling faults, stuck-open access transistors, write-recovery faults, destructive reads, and others).

The memory BIST controller also supports a *Checkerboard* algorithm to test DRAMs and detect coupling faults based on a physical map. The *Checkerboard* algorithm can also be combined with the *March* algorithm to test RAM retention time.

A patented shadow read test scheme for multiple-port memories yields superior quality. This shadow read test covers bit-line shorts and word-line shorts that belong to different ports.

The controller also supports a two-pass, read-only algorithm to test ROMs. The two-pass algorithm eliminates possible error-masking for higher coverage.

The memory BIST controller is fully compatible with scan and logic BIST so that the few internal memory BIST controller faults not tested during the memory BIST operation can be covered using the scan or logic BIST test.

The memory BIST collar also offers a significant number of features that simplify testing user logic connected to the memory. For example, observation points can be added at the memory inputs, and memory outputs can be controlled from scannable registers. The collar also supports a number of modes for bypass or write-through of the memory for scan test.

Diagnosis

Several features of LogicVision's memory BIST are available for diagnosing failures. The DONE output of the controller is used to indicate whether the failure occurred in the memory itself or the controller. When multiple memories are tested in parallel, a separate *go/no-go* signal can be generated for each one in addition to the global *go/no-go* signal to locate any failing memories. Optional comparator status outputs can be monitored in real time to log each compare error that occurred. The diagnostic resolution is adjustable down to the bit level.

Another memory diagnostic capability is a stop-on-n^{th}-error feature that can stop the controller after the nth miscompare and scan out all the relevant information on the failure through a serial interface—typically the IEEE 1149.1 TAP interface. The BIST controller can then resume testing and stop on the next error.

Legacy Core Test

icBIST provides the hierarchical capability that supports testing *legacy cores* embedded within an SOC device. A legacy core is any predesigned functional block that must normally be made accessible from chip pins in order to apply existing, core-specific tests.

Core-based design has become a necessity with the advent of VDSM technology. Many cores are available to chip designers from various sources. The cores can be classified as follows:

- *Soft* (synthesizable RTL)
- *Firm* (gate-level netlist)
- *Hard* (non-modifiable layouts, often called *legacy cores*)

The BIST options to icBIST enable testing of most soft and firm cores because these cores can be modified to add testability and merged with the rest of the user logic. Many legacy cores have adequate testability so that they can be transparently merged with the rest of the logic. Such legacy cores have existing scan chains that can be connected to a logic BIST controller.

However, some legacy cores do not meet scan and logic BIST design rules. New solutions are required to test these cores. Another issue that must be addressed by any solution is the testability of the logic surrounding the legacy cores. It is very difficult to propagate faults through untestable cores. Core outputs must be controllable and inputs must provide observation for the logic that drives the core.

LogicVision's solution provides automatic generation of a configurable *collar* for isolation and access of a core. This feature supports hard, firm, and soft cores, provides dedicated isolation and access on a per-pin basis, and allows observation and control of the surrounding logic. Additionally, this solution provides automated assembly of the *collared core* into the chip-level design. The assembly generates I/O multiplexing to provide external access for test and allows control of legacy core test via the IEEE 1149.1 TAP. Finally, LogicVision's solution provides automatic generation of a testbench that invokes the existing legacy core test.

The ability to access and test deeply embedded cores independently of each other and of other, surrounding functions improves both testing and diagnosing the core while minimizing the defect escape rate. Direct pin access to an embedded core can also be useful for verification or conformance testing of the core function.

Operation

The LogicVision Legacy Core Test option addresses legacy-core testing through an architecture that surrounds the core with an isolation-and-access test socket. There are three modes of operation:

● *Normal* operational mode—the collar passes the functional signals between the core and surrounding logic without altering them.

● *Test isolation* operational mode—the collar serves as an isolation mechanism to facilitate testing the surrounding core's logic.

● *Test access* operational mode—the collar connects the I/Os of the legacy core with a direct access path to external chip pins. This access path allows the application of pre-existing test and diagnostic patterns.

Selecting operational modes is controlled either directly from pins of the chip or through an IEEE 1149.1 TAP on the chip. To facilitate testing multiple legacy cores in parallel or sequentially, test collars can be individually selected and controlled.

The testing of very large chips that require using multiple BIST controllers distributed in several cores is described in Chapter 5, "Hierarchical Core Test."

Phase-Locked Loop Test

LogicVision also provides an optional solution for at-speed testing of PLLs. The pllBIST option automatically generates synthesizable RTL code, synthesis scripts, and verification and test patterns. pllBIST enables at-speed testing of PLLs using a standard digital tester.

The custom-generated pllBIST Verilog or VHDL design object connects to only the inputs and outputs of the PLL to test. No changes or connections to the internal nodes of the PLL are necessary. A digital multiplexer drives the input of the PLL. Testing the PLL is fully synchronous, making it suitable for very high-speed tests. pllBIST tests PLLs that contain phase-frequency detectors (PFDs) and charge pumps or low pass filters.

pllBIST measures the loop gain, frequency lock range (upper and lower values), lock time, and jitter (relative to PLL input) one parameter at a time. pllBIST then compares the measured values on chip against pass/fail limits. A tester loads these limits serially. The binary-encoded measurement results are serially shifted out, together with a pass/fail bit for each limit.

pllBIST measures the peak-to-peak and/or RMS jitter. The *cumulative distribution function* (CDF) limits for the jitter (0 to 100 percent for peak-to-peak; 16 to 84 percent for RMS) are loaded in at test time from off chip or are hard-wired on chip. By loading in various limits, the entire CDF can be measured for rapid, low-cost characterization.

pllBIST implements an all-digital BIST circuit that fully tests PLLs at application speed. The benefits of pllBIST include the following:

● Measures loop gain, lock range, lock time, and jitter, then compares to binary-encoded limits

  ● Improves quality—complete at-speed test at wafer, package, or board level

  ● Reduces test costs—faster test; lower cost tester

  ● ATE supplies constant frequency reference and low-speed control; for example, through an 1149.1 TAP

● Provides all-digital circuit that is synthesizable from Verilog and VHDL

  ● Reduces time to market by facilitating layout automation and ATPG

  ● Requires no changes to PLL and no connections to analog nodes

● Features diagnostic output and pass/fail result that helps debug design and yield problems immediately after receiving first silicon

Analog-to-Digital Test

LogicVision also provides an optional solution for at-speed testing of ADCs. The adcBIST option automatically generates synthesizable RTL code, synthesis scripts, and test and verification patterns. adcBIST enables at-speed testing of ADCs using a standard digital tester.

The custom-generated adcBIST Verilog or VHDL code connects to only the inputs and outputs of the ADC to test; no changes or connections to the internal nodes of the ADC are needed. The analog input is driven by an analog multiplexer or driven directly by a tri-state digital buffer in series with a resistor. (If the ADC has a low input impedance, analog buffering is necessary.) Operation is fully synchronous with the sampling clock, making it suitable for low- or high-speed operation. Multiple ADCs can be tested serially.

adcBIST generates an analog stimulus for connection to the ADC via an RC filter (RC within +/-20 percent of the value specified by the automation software), which may be on chip, off chip, or both, depending on sample rate, number of bits, RC linearity, and ADC input impedance. The output of the ADC, as it samples the stimulus, is processed by adcBIST circuitry using a patented technique. Five measurement results are generated. Offset, gain, second and third harmonic distortion, and idle channel noise are compared on chip against pass/fail limits serially loaded in from the tester. The binary-encoded measurement results are serially shifted out, together with a pass/fail bit for each limit. Test time can be traded off against accuracy, if necessary, and a settling time can be programmed for AC-coupled connections.

The benefits of adcBIST include the following:

● Measures key specifications and compares to binary-encoded limits on chip
  ● Reduces test costs: lower cost, digital tester
  ● ATE supplies only low-speed control; for example, through 1149.1 TAP
  ● Improves quality for high-speed ADCs: complete at-speed test at wafer, package, or board level
● All-digital circuit that is synthesizable from Verilog or VHDL
  ● Reduces time to market by facilitating automation and ATPG
  ● Requires no changes to ADC
● Diagnostic output and pass/fail result helps debug design and yield problems immediately after receiving first silicon

Fault Insertion

LogicVision also provides an optional solution for system-level fault insertion. Fault insertion has been used for many years to evaluate the effectiveness of system diagnostic software. Even though it is possible to perform this evaluation using simulation, it is usually difficult because of the absence of proper tools and models.

LogicVision provides a unique infrastructure based on IEEE 1149.1 Boundary Scan to implement fault insertion in an optimal manner. Designers can leverage existing BScan cells to inject faults, while the system is running in functional mode, at selected pins of an integrated circuit. Faults can also be injected at selected internal nodes using scan cells. Faults can emulate stuck-at 0 or 1 conditions. Multiple faults can be injected in several chips or within the same chip at the same time.

The duration of the fault conditions are under user software control so that faults can be applied for very long or very short periods of time or can emulate intermittent faults. The faults can be made *persistent*. These faults remain in effect even if the Bscan chain is used by the system software diagnostics program to apply off-line structural tests. The offline tests locate more precisely the fault that was detected by online detectors.

The fault insertion circuitry is generated and connected automatically to the rest of the chip at the same time as the TAP and Bscan circuitry. The location of the faults is documented using the Boundary Scan Description Language (BSDL), which supports the IEEE1149.1 standard. Existing boundary-scan cell layouts can be reused to implement a complete fault insertion cell.

Fault insertion provides these important benefits:

● High quality of system software diagnostics

   Software can be tested thoroughly in a real system environment to determine the accuracy of diagnostics and the fault tolerance of the system. New versions of software can be regression-tested on any hardware system configuration. It can even be tested at the customer site.

● Faster time to market

   Hardware and software system integration and debugging are greatly accelerated by the use of fault insertion. The large number of potential fault sites test several faulty conditions and quickly isolate each hardware or software component that is not responding appropriately.

External Memory Test

LogicVision also provides an optional solution for at-speed test and diagnosis of memories at the board or system level.

The memBIST-XT capability provides a complete, automated BIST solution for off-chip memory structures in hierarchical assemblies. This includes memories on printed circuits boards (PCBs) and multi-chip modules (MCMs), as well as other hierarchical structures. The memBIST-XT controller can test clusters of memory such as stand-alone memory chips, Single Inline Memory Modules (SIMMS) and Dual Inline Memory Modules (DIMMs). The controller is embedded in the chip or chips that interface with the memory array.

The memBIST-XT controller is responsible for executing the at-speed, self-test operation that includes the generation of test data and expected responses, and the capture and compare of actual responses. It can be controlled either from an 1149.1-based TAP controller on the board or directly through the pins on the chip. The controller can account for empty memory slots while testing a memory array.

The BIST interface provides the interface between the controller and the external memory structure. It includes the multiplexing logic for switching between functional signals and controller signals. The multiplexing can be done before the last stage of flip-flops before the pads to avoid affecting the functional timing.

memBIST-XT provides automation tools for generating the controller and interface based on user input. memBIST-XT processes user-specified design information about the target memory cluster, BIST configuration information, and special options.

Based on this information, memBIST-XT automatically generates the following:

● Controller and interface IP design objects in synthesizable Verilog or VHDL descriptions

● A testbench (Verilog or VHDL) to verify the functionality of BIST structures in simulation

● Synthesis scripts that target a single ASIC or FPGA

The memBIST-XT controller implements variations of March-like algorithms with programmable read-and-write access sequences. This programmability allows the controller to interface to a wide variety of memory types (SRAM, DRAMs, and SDRAMs). The algorithms test for stuck-at faults, cell-related faults, shorts, and coupling faults for bit lines and data and address buses.

For diagnosis, memBIST-XT has a stop-on-failure option to allow scanning out the failing address and data values through the controller and interface before resuming the test. It also supports the generation of a map of failed interconnects and a map of failing memory chips within an address range.

Benefits of the memBIST-XT test approach include the following:

● Use of test speeds several orders of magnitude faster than traditional approaches that rely on the use of a tester or embedded processor to test memory

● Higher quality products because of the use of proven algorithms and patterns applied at speed

● Reduced time to market because the generation of BIST circuitry and testbenches is automated

● Reduced debug time through the use of diagnostic features that quickly identify the failing memory chips and/or interconnect

At-Speed Interconnect Test

LogicVision also provides an optional solution for ASI tests at the board or system level. The TAP and boundary-scan technology described earlier in this chapter allows the application of low-speed structural tests to board or system interconnect. This method detects manufacturing defects, such as shorts and opens, and makes them easy to diagnose and repair whenever possible. However, manufacturing defects that manifest themselves as delay faults cannot be detected.

Many companies today perform interconnect testing by using In-Circuit Test (ICT) techniques. Testers that provide this capability typically contain hundreds to thousands of physical probes (often called a *bed of nails*) that are used to contact chip pins and board testpoints to perform the interconnect tests. For today's designs, however, the use of ICT is often limited because of the total number of probes required and the physical limitations of accessing chip pins because of packaging techniques such as BGAs, as well as overall board densities. In addition, testing for interconnect delay faults is not practical because the tester hardware cannot apply at-speed signals to hundreds or thousands of probes.

The method most commonly used to achieve at-speed interconnect testing is functional testing at the system level. However, when an error is detected using this approach it can take days or weeks to diagnose where the problem is located. Clearly, a better method of testing interconnect at application speeds is needed to ensure that today's highly complex systems can be introduced at the appropriate times and that production deliveries and cost can be kept under control.

LogicVision provides a patented jtag-XLi option that enables designers to apply and capture data across board interconnect in one system clock cycle. This technology builds upon the standard IEEE 1149.1 JTAG board test approach. jtag-XLi uses a combination of new at-speed boundary-scan cells and an ASI test controller block to drive necessary timing signals to these new cells.

The at-speed boundary-scan cells consist of standard 1149.1 cells augmented with specialized logic and a system-clocked flip-flop. The cells are still scanned directly from the TAP using the TCK test clock. However, instead of being controlled by the standard Update and Capture signals generated by the TAP controller, the at-speed boundary-scan cells are controlled by at-speed versions of these signals generated by the embedded ASI test controller. A board-level synchronization signal is also routed between the ASI test controllers residing on the chips involved in the at-speed interconnect test. This signal ensures that the operation of the two or more ASI controllers are all aligned to the same system clock cycle.

jtag-XLi includes automation software to create customized RTL descriptions of the enhanced boundary-scan cells and ASI controller in Verilog or VHDL format. It also creates and assembles a standard IEEE 1149.1 TAP, the at-speed boundary-scan cells, and the ASI test controller into a design.

One of the key advantages of the jtag-XLi approach is that it is completely transparent to IEEE 1149.1-based testing. Indeed, a board design can include a mixture of chips that incorporate the jtag-XLi technology with chips simply compliant with the IEEE 1149.1 standard. Any existing boundary-scan test hardware, test and diagnostic patterns, and software tools can still be used with both chip types without modification. In this scenario, only the interconnect between chips that incorporate the jtag-XLi technology is tested at speed. The remaining interconnect is tested at standard TCK rates.

References

[1] *IEEE Standard Test Access Port and Boundary-Scan Architecture* (IEEE Standard 1149.1-1990 with IEEE Standard 1149.1a-1993).

[2] *Supplement to IEEE Standard 1149.1-1990, IEEE Standard Test Access Port and Boundary-Scan Architecture* (IEEE Std 1149.1b-1994).

[3] Semiconductor Industry Association. *National Technology Roadmap for Semiconductors*, 1997.

[4] M. Abramovici, M. Breuer, and A. Friedman, *Digital Systems Testing and Testable Design*. IEEE Press, 1995.

[5] P. Franco et al, "Analysis and Detection Of Timing Failures in an Experimental Test Chip," in *Proceedings of the 1996 International Test Conference*, Washington, DC, October 20-25, 1996.

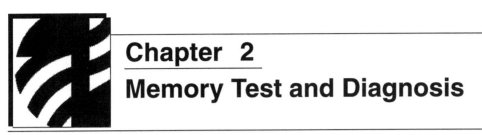

Chapter 2
Memory Test and Diagnosis

This chapter introduces the concept of embedded test for embedded memories.

The increased use of embedded memories has resulted in an urgent need to develop ways of adequately testing memories that are inherently difficult to access. This chapter discusses some of the techniques that have been used to solve this testing problem. Important difficulties associated with these techniques are outlined, leading to the conclusion that a better approach is needed. The BIST approach for embedded memories, and its advantages, is described, showing how BIST can be used to effectively test any number of embedded memories integrated into a single chip.

Overview

A memory is considered *embedded* if some or all of its terminals are not directly connected to the pins of the host chip. Figure 2-1 shows a simplified representation of an embedded memory.

Figure 2-1 Representation of an Embedded Memory

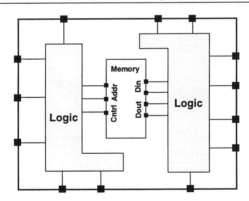

Embedded memories are becoming increasingly common in large ASIC designs. It is not unusual to have multiple embedded memories in a single ASIC. Not only are embedded SRAMs being used, but ROMs and DRAMs are being integrated into VDSM ASICs.

Difficulties in Testing Embedded Memories

In general, testing any memory requires applying a large number of test patterns to its address and data inputs and observing the values read out from its data outputs. These test patterns typically need to be applied at the speed required in the intended application in order to properly stress the memory and detect any timing-related defects. Embedded memories provide a unique testing challenge because they lack direct accessibility. A method is needed to apply the necessary test patterns to the embedded memory's inputs and observe its outputs. Figure 2-2 illustrates two techniques aimed at achieving these testing requirements.

The first technique is perhaps the most direct and entails routing all of the embedded memory's inputs and outputs directly to the host chip's pins. Multiplexing logic is used to choose between this routing during test mode and the normal functional paths. However, there are three main problems with this approach:

● The amount of extra interconnect needed to accomplish this routing can become prohibitive. Interconnect is becoming increasingly important in today's designs.

● Even if the extra interconnect can be tolerated, the resulting interconnect typically must be routed over large portions of the chip, leading to large and probably uneven propagation delay. This condition makes applying the test patterns at speed difficult if not impossible.

● The number of memory inputs and outputs might be larger than the number of chip-level signal pins.

The problems become worse when there is more than one embedded memory on a chip. Indeed, chips that contain dozens and even hundreds of embedded memories are not uncommon.

Figure 2-2 Direct Access for an Embedded Memory

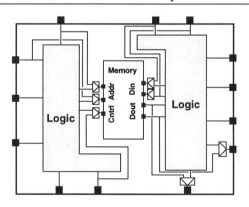

The second technique, shown in Figure 2-3 on page 37, involves placing a scan chain around the embedded memory; that is, a scan flip-flop is placed next to each memory input and output. This technique is similar to the concept of boundary scan at the chip level.

The added scan chain provides the capability of scanning in test patterns and scanning out test responses. This capability solves the three problems associated with the first technique but results in two other problems:

- Because the test patterns and responses need to be scanned in and out, the test application time can grow by orders of magnitude. Instead of applying a test pattern every clock cycle, n clock cycles are needed to scan each pattern in and out, where n is the length of the scan chain. At a minimum, n is equal to the number of the memory's inputs and outputs. Even for medium-sized embedded memories, n can equal 100 or more.

- Because the test patterns cannot be applied to the embedded memory on a cycle-by-cycle basis, an at-speed test of the memory cannot be performed.

Figure 2-3 Scan Access for an Embedded Memory

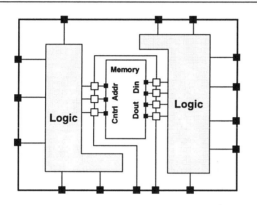

It is clear from the preceding arguments that the two techniques are inadequate in most cases. Both techniques are aimed at providing access to the embedded memory's inputs and outputs from the host chip's pins. Removing this access requirement greatly simplifies the testing problem.

BIST for Embedded Memories

The key to solving the embedded memory input and output access problem, described in the section "Difficulties in Testing Embedded Memories" on page 36, is to generate the necessary test patterns and analyze the test responses local to the embedded memory. This, in essence, is the memory BIST approach, as illustrated in Figure 2-4.

Figure 2-4 BIST Approach for an Embedded Memory

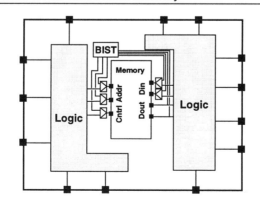

A local BIST controller is used to generate the test patterns and observe the test responses from the memory. Multiplexing logic chooses between address and data originating from the BIST controller or from the system logic.

Figure 2-5 provides a more detailed view of a typical embedded memory BIST architecture.

Figure 2-5 Typical BIST Architecture for Testing Embedded Memories

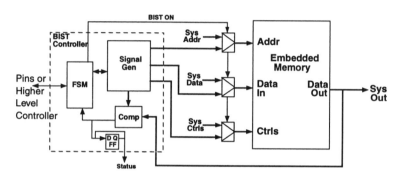

The BIST controller consists of three main blocks:

- A *signal generation block* (Signal Gen) contains the circuitry to produce the address, data, and control values (Sys Addr, Sys Data, and Sys Ctrls, respectively) necessary to create each test pattern that is to be applied to the memory. This block typically contains an up/down counter for generating the address sequences needed by most memory test algorithms.

- A *comparator* (Comp) compares the values read out of the memory with expected values generated by the signal generation block on a cycle-by-cycle basis. The result of each comparison is accumulated into a status flip-flop (FF) in order to provide a go/no-go result at the end of the test. Often, the comparison result is brought out to a chip pin for real-time monitoring.

- A *finite state machine* (FSM) is used to control the overall sequence of events. For example, it determines if the address counter should be counting up or down or if the data being generated should be a marching 0 or marching 1 pattern.

The BIST controller typically requires very simple initialization and control to operate. For example, the initialization might indicate which memory test algorithm to apply or simply clear the status flip-flop. The initialization and control might come from a chip-level test controller or from a source external to the chip. In many cases, the controller is wired to an IEEE 1149.1 TAP, which provides an industry-standard interface for accessing the memory BIST controller.

Generally, a single BIST controller can be used to test more than one embedded memory on a chip. Figure 2-6 on page 39 illustrates a typical scenario for a chip.

Figure 2-6 Sharing a BIST Controller Among Several Embedded Memories

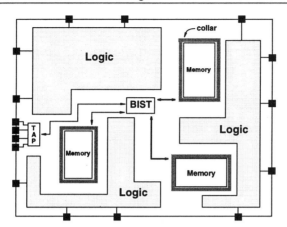

Figure 2-6 shows a single controller (BIST) used to test three separate embedded memories. Multiple memories can be tested either sequentially or in parallel. The advantage of testing the memories in parallel is a reduced test time. There are, however, disadvantages to parallel testing:

- The power consumption that results from testing several memories together can be high.
- Certain BIST controller resources must be duplicated. For example, a separate comparator is needed for every memory tested in parallel.

There is also a potential disadvantage to using a single controller for multiple embedded memories. If the memories are not localized to one area of the chip, a large amount of wiring might be needed to route the address and data lines from the controller to each of the memories. This disadvantage is especially true for data lines, given the large data widths (64 bits, 128 bits, and so on) many memories currently have.

memBIST-IC Approach

To solve this problem, LogicVision's memory BIST approach requires that only one data line be routed to each embedded memory, regardless of the memory's data width. This approach is referred to as a *serial memory BIST,* because the data values for each test pattern are provided serially to the memory. Figure 2-7 on page 40 shows this approach.

The serial memory BIST approach involves using the memory itself to shift the incoming data bits across the data width of the memory. A multiplexer is added to each data input so that, in BIST mode, the value written into data input pin k consists of the value read out of data output port k-1.This operation results in each consecutive data pattern being a 1-bit shifted version of the preceding pattern.

Figure 2-7 LogicVision's Patented Serial BIST Approach for Embedded Memories

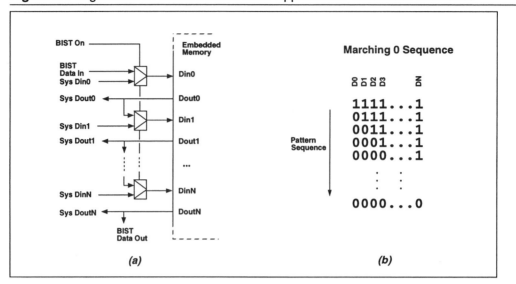

(a)   *(b)*

Although this might seem like a limitation on the test algorithms that can be applied to the memory, most common test algorithms use *marching* test pattern sequences. A marching sequence consists of consecutively writing a 0 or 1 value across the width of the data word. For example, a marching 0 pattern sequence on a background of logic 1 values is illustrated in Figure 2-7. Notice that each consecutive pattern is a 1-bit shift of the previous one. This type of pattern is useful for detecting coupling faults within a memory word without knowing the physical structure of the memory. The coupling faults are usually covered by repeating a march test with different foreground and background patterns.

The serial memory BIST approach allows the design engineer to trade off test time for area by dividing up the word in slices of equal size. In fact, the design engineer can revert to a fully parallel approach if it makes sense for the application. This trade-off is explained in more detail in the section "Embedded Test for Multiple Memories" on page 46.

Description of Controllers

This section introduces the memory BIST architecture, the various architectural components, the supported memory test configurations, and the memory BIST operating protocol.

Memory BIST Architecture Overview

The block diagram of BIST-enabled memories in Figure 2-8 on page 42 illustrates the basic memory BIST architecture. All tests are applied to the memories from the memory BIST controller using a memory I/O interface circuit called a memory BIST *collar*. A single controller can be used regardless of the number of memories tested. However, multiple controllers can be implemented to optimize test time and routing congestion. The memory BIST controller can support a number of different diagnostic options including a pass/fail flag, a serial interface for shifting out failure data, and a parallel interface with a failure flag per bit.

- The serial interface works with a *stop-on-error* feature that allows data that characterizes each failure in each memory to be captured and shifted out. This approach is useful for debugging RAMs with design flaws during prototype testing.

- The parallel interface is useful in cases where defective RAMs are bit-mapped either for yield improvement or for identifying the potential for laser repair.

The LogicVision memory BIST architecture uses a patented serial-access technique to provide a simple communication interface between the controller and the data path of the memory under test. This serial linkage minimizes the routing and gate overhead associated with memory BIST and offers greater versatility in testing several memories with a single memory BIST controller.

This serial-data architecture is especially well suited for memories that have relatively wide data paths, where minimizing routing area and congestion are of concern. The memory BIST controller needs to provide only one bit of the test data-in to each memory that it tests. Additionally, the memory BIST controller must observe only one output bit from each memory that it tests. More output bits can be observed to reduce test time at the expense of more area.

These two features significantly reduce the amount of routing between the memories and the memory BIST controller compared to conventional memory BIST schemes.

Figure 2-8 High-Level View of the Memory BIST Architecture

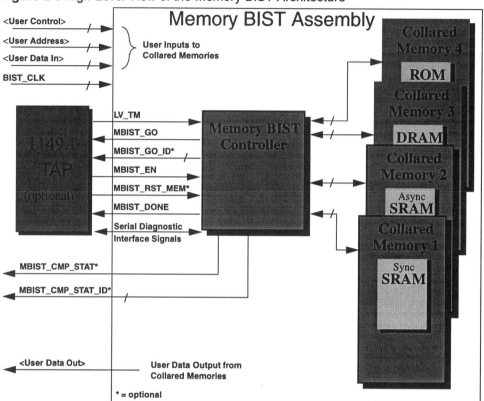

Memory BIST Controller

The main component of the memory BIST architecture is the memory BIST controller, which manages the sequencing of control signals, address, and data to the collared memories. The memory BIST controller consists of several functional modules, as shown in Figure 2-9 and described below.

- Memory Counter—counts and controls the sequencing of memories when there are multiple test steps to execute.

- Port Counter—counts and controls the sequencing of various ports of multi-port memories.

- Address Counter—counts through the address space of the memory to test according to the specified algorithm.

- Bit Counter—counts and controls the sequencing of the test data to a bit slice.

- Timing and Sequencing—controls the overall timing and sequencing of control signals necessary to access the memories. This module also controls the interaction of the other modules within the memory BIST controller.

- Comparator—contains one or more comparators used for checking the actual data from the memory against the expected data generated by the memory BIST controller.

- Algorithm—contains the hardware implementation of the selected algorithms necessary for the memory BIST configuration.

Figure 2-9 BIST Controller Functional Diagram

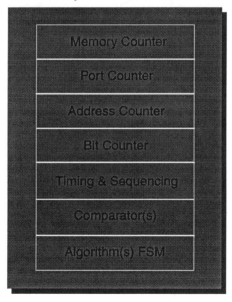

daisyChain = ON   Memory BIST Controller

Memory Counter

Port Counter

Address Counter

Bit Counter

Timing & Sequencing

Comparator(s)

Algorithm(s) FSM

Collar

The second component of the memory BIST architecture is the memory collar. The collar provides the interface between the memory BIST controller and the memory. Figure 2-10 shows how a RAM is connected to the collar. The collar consists essentially of multiplexing logic that chooses between signals generated by the memory BIST controller and signals from the user logic. The collar might also contain scan logic to isolate the memory from the surrounding logic during scan-based testing.

Figure 2-10 Connections to the Collar for a RAM

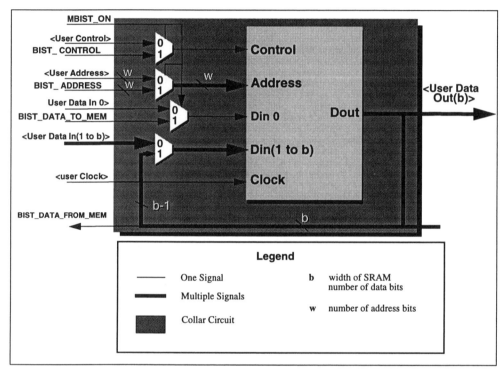

Figure 2-11 on page 45 shows how a ROM is connected to the collar. In addition to the multiplexing logic, a ROM collar also contains a MISR to compress the ROM output values into a signature.

Figure 2-11 Connections to the Collar for ROM

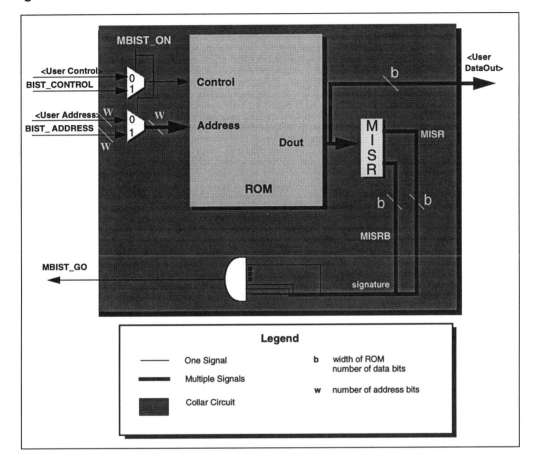

BIST Operating Protocol

Figure 2-12 on page 46 illustrates the protocol used by LogicVision's memory BIST controller. The input pin *MBIST_EN* initiates the test. The first falling edge of *MBIST_CMP_STAT* after the rising edge of *MBIST_EN* marks the first clock cycle of the test. If an error is discovered, the *MBIST_CMP_STAT* signal goes low for the next cycle and then returns to high. *MBIST_GO* falls and remains low for the remainder of the test. The rising edge of *MBIST_DONE* marks the end of the test. If *MBIST_GO* is still high at that time, all memories have passed the test.

Figure 2-12 BIST Operating Protocol

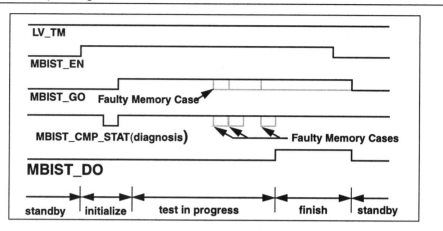

It is not necessary to continue the test if the *MBIST_GO* signal has fallen. However, if the *MBIST_CMP_STAT* signals are being monitored, testing should be continued to find more memory failures. It is also important to verify that *MBIST_DONE* rises at the appropriate time to rule out a manufacturing defect within the memory BIST circuitry. Whether a failure occurs or not, after the *MBIST_DONE* rises, *MBIST_EN* can be deasserted to stop the test and return to functional mode.

LV_TM, when driven high, is used to set up the collar and memory BIST controller for scan or logicBIST.

When a TAP is present on the chip, a specific instruction bit can be used to initiate the memory BIST. In addition, status bits can be used to sample the *MBIST_GO* and *MBIST_DONE* signals.

Embedded Test for Multiple Memories

As mentioned in the section "Memory BIST Architecture Overview" on page 41, a single memory BIST controller can test any number of memories. Memories can be tested serially or in parallel. The manner in which the memory BIST controller sequences through the memories is referred to as a *test configuration*. The configuration is thus a set of sequential steps of independent tests. Each memory to test is assigned to a particular step. The memories are organized within each step to achieve an optimal configuration.

Sequential Testing

Grouping memories into separate steps is called *sequential testing*. All memories in step 1 are tested before memories in step 2, and so on. The advantage of testing memories sequentially is that the output comparators can be shared among all memories. The total number of comparators required within the memory BIST controller is equal to the greatest number required by any single memory. The memory BIST architecture that supports sequential testing is illustrated in Figure 2-13 on page 47.

Figure 2-13 Sequential Testing of Multiple Memories

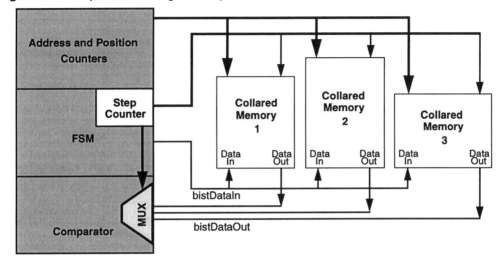

Parallel Testing

Grouping memories into a single step is referred to as *parallel testing*. The advantage of parallel testing is that test time is reduced because all of the memories within the step are tested at the same time. The disadvantage is that the comparators cannot be shared, resulting in a greater overhead than the serial testing approach. Figure 2-14 illustrates the memory BIST architecture that supports parallel testing.

Figure 2-14 Multiple RAMs in a Parallel Configuration

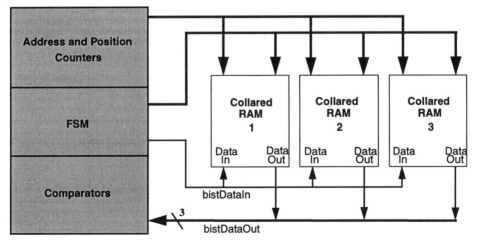

Using a parallel test configuration, all memories receive the same input data. If the LogicVision serial data-in feature is used, a single data-in signal, *bistDataIn,* is connected to all of the memories in parallel. The most significant bit of each memory (*bistDataOut*) connects to a separate comparator inside the memory BIST controller. (For the example in Figure 2-14, three comparators are required.)

Parallel Configuration of a Single RAM

A single RAM can be tested in parallel slices by assigning a separate comparator per slice. The parallel test configuration reduces the test time. Every comparator is driven by a single data-out signal from the memory. However, adding extra comparators to test a single RAM in parallel increases routing congestion because of the extra data-out signals.

Figure 2-15 illustrates the memory BIST architecture that supports parallel testing of a single RAM. One-bit slices are used to create a full parallel test.

Figure 2-15 A Single RAM in a Parallel Configuration

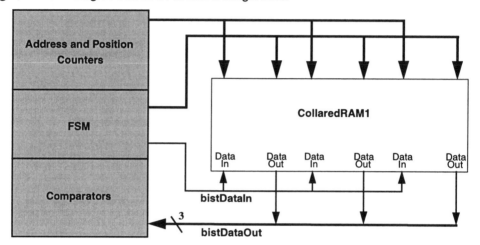

Chained-Memory Configuration

The chained-memory test configuration consists of several RAMs wired in series to be tested as a wider RAM. The primary benefit of the chained-memory configuration is that it provides flexibility to combine narrow RAMs to create a slice width that allows them to be tested in parallel with wider RAMs.

Data from the memory BIST controller goes to the first memory in the chain. The output of each subsequent memory in the chain is connected to the input of the next memory until the last memory in the chain returns the data-out signal to the comparator in the memory BIST controller.

Figure 2-16 on page 49 illustrates the architecture that supports the chained-memory configuration for three RAMs.

Figure 2-16 Chained-Memory Configuration for Three RAMs

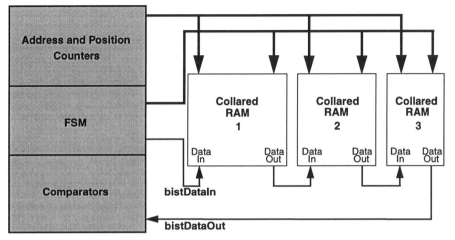

Step Compatibility Rules

There is no limit to the number of memories that can be tested in the same step provided that these compatibility rules are satisfied:

● All memories must use the same algorithm.

● All memories must use the same operation set (Read waveform, Write waveform, and so forth).

● All memories must have the same number of bits per data slice. For chained memories tested in parallel with other memories, the sum of bits for all of the memories in the chain must equal the data slice width.

Creating an Optimal Test Configuration

In this section, a combination of sequential, parallel, and chained-memory test configurations is illustrated. Figure 2-17 on page 50 illustrates a memory BIST architecture that supports a test configuration for five RAMs:

● A 32-bit x 8-word SRAM

● A 16-bit x 16-word SRAM

● Two 8-bit x 8-word SRAMs

● A 32-bit x 16-word DRAM

The example assumes that the SRAMs are equivalent except for size. Because the SRAMs and DRAM are tested with different algorithms, they must be tested in separate steps. The 16-word SRAM can be tested in the same step as the 8-word SRAMs.

Looking at the width of the SRAMs in the first step, it is possible to determine a common width at which the majority of SRAMs can be tested in parallel. This example uses a common width of 16 bits. To achieve this common width, the 8x32 SRAM is divided into two 8x16 sections and the two 8x8 SRAMs are chained to create an 8x16 SRAM. Now the two chained 8x8 SRAMs, the 8x32 SRAM, and the 16x16 SRAM can be tested in parallel.

The second step tests the 16x32 DRAM and can reuse any number of the comparators from the first step. Depending on total test time constraints and diagnostic requirements, other configurations can be used.

There is no deterministic algorithm for creating an optimal test configuration for a design. Balancing the benefits and costs of the alternative configurations achieves the testing goals of reduced testing time, reduced physical overhead, and increased diagnostic resolution. Such trade-offs depend on the relative importance of each of the goals.

Figure 2-17 Combination of Basic Test Configurations for Testing Multiple SRAMs

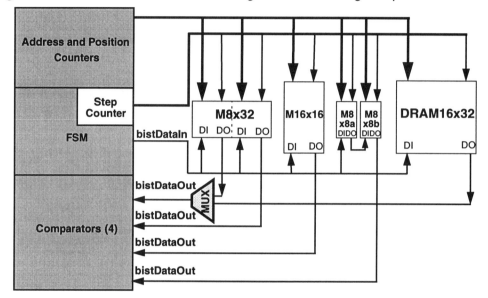

Note: This figure shows the mux outside the comparators; in actuality, this mux is within the comparator.

Reducing Test Time

The key to reducing test time is the number of comparators. Every comparator added to a configuration reduces the testing time. The shortest possible test starts with the minimum number of steps and puts a comparator on every data bit.

Testing time for each step is roughly proportional to the number of comparators multiplied by the slice width. This is not the number of clock cycles—it is a number that can be used to compare alternative configurations.

For example, a pair of 8K by 8 memories configured in a chain takes approximately the same amount of time to test as a configuration where the memories are tested in separate steps (1 comparator x 16 bits = 1 comparator x 8 bits + 1 comparator x 8 bits). Using LogicVision's *SMarch* algorithm as an example, the chained-memory configuration takes 4.2 million clocks and the two-step configuration takes 4.5 million clocks. Using the same number of comparators, the configuration with more steps takes a little longer than a more parallel configuration.

Reducing Physical Overhead

Using more comparators results in more logic overhead. The number of comparators in the memory BIST controller is equal to the maximum number of comparators in any step. If three comparators are used in the first step and two comparators in the second step, there will be three comparators in the memory BIST controller.

When the memories in the second step are tested, two of the three comparators from the first step are reused. A multiplexer is added to each comparator input the first time it is reused and an extra multi-plexer input is added every time it is reused after that.

Every memory has at least one serial data output signal that usually goes to the memory BIST con-troller but alternatively goes to another memory in the chained-memory configuration. Every additional comparator on a memory adds another wire running from the memory to the memory BIST controller. These extra wires contribute to routing congestion.

Increasing Diagnostic Resolution

Diagnostics are recorded on a per-comparator basis. When a failure occurs, the failing address and bit slice is known, but not necessarily which bit within that slice. Adding more comparators increases the bit diagnostic resolution.

Diagnosis

This section discusses two diagnostic approaches: a standard method that uses parallel compare sta-tus lines and a *stop-on-nth-error* method.

One of the most basic diagnostic capabilities supported by the memory BIST methodology is illus-trated in Figure 2-18. (Although an IEEE 1149.1 TAP is shown as the interface to the BIST controller, a direct pin interface can also be used.) In this scenario, one or more *compare status* sig-nals from the BIST controller are routed directly to chip pins. These compare status signals provide cycle-by-cycle compare results. A compare status output can be made to correspond to anything from the full memory word to a single bit of the word. Each of these outputs are routed directly to pins for monitoring by the tester.

Figure 2-18 Standard Diagnostic Approach Using Parallel Compare Status Lines

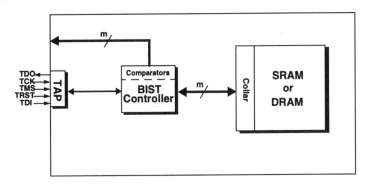

To extract enough failure data to resolve down to individual bit failures (typically referred to as *full failure bit mapping*), as many compare status pins as bits in the memory word are needed. However, this can result in an unacceptably large number of pins. Furthermore, the tester needs to operate at the same clock rate as the BIST controller, because the compare status pins provide new data at every clock cycle. Fortunately, most problems can be diagnosed at speeds that are appropriate for low-performance testers. However, some problems might require running the test at speed.

To eliminate the above constraints, a purely scan-based diagnostic methodology is also supported, as Figure 2-19 illustrates. This approach is referred to as *stop-on-nth-error* because the BIST controller is equipped with an error count register that is scan-initialized before each run. When the controller has encountered a number of errors equal to that specified in the error register, it stops so that all pertinent error information (algorithm step, address, failing bit position, and so forth) can be scanned out of the controller. By iteratively scanning in subsequent error counts and running the controller, all bit-fail data can eventually be extracted.

Figure 2-19 Diagnostic Approach Using Stop-On-Nth-Error Serial Scan

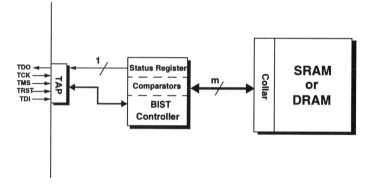

Compare Status Pins Approach

The compare status pins diagnostic approach involves having the tester directly monitor one or more compare status signals routed from the memory BIST controller directly to chip pins. A global compare status (*CMP_STAT*) signal can be used if resolution down to only the failing word is required. If bit-level resolution is needed, individual compare status (*CMP_STAT_ID*) signals for each comparator must be used.

The number of chip pins needed to support the *CMP_STAT_ID* signals depends on the following:

● If a memory is configured to be tested in parallel bit slices, each bit slice will have its own comparator and generate a separate *CMP_STAT_ID* signal. Figure 2-20 on page 53 shows a simple example in which a memory is divided into three 2-bit wide slices. Each of these slices requires a *CMP_STAT_ID* pin.

● The *CMP_STAT_ID* signals of different memories tested in parallel by the controller, that is, memories that are tested in the same step, must be routed to separate chip pins. The *CMP_STAT_ID* signals of different memories tested sequentially (in different steps) are automatically multiplexed inside the memory BIST controller. Therefore, the number of *CMP_STAT_ID* pins is equal to the maximum number of comparators in any step.

The compare status pins diagnostic approach has the following requirements:

● Enough chip pins are available for routing all of the required *CMP_STAT_ID* signals.

● The tester can monitor all of the *CMP_STAT_ID* pins at the memory BIST controller's clock frequency.

When either of the above two requirements cannot be met, the stop-on-nth-error scan-based diagnostic approach can be used.

Figure 2-20 Memory Requiring Multiple Individual Compare Status Signals

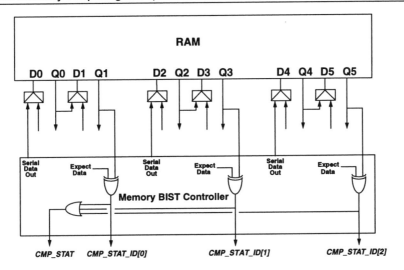

Stop-On-Nth-Error Approach

The stop-on-nth-error diagnostic approach eliminates the need to monitor status pins by scanning out all pertinent failure data from the controller's internal registers. Figure 2-21 shows a typical LogicVision memory BIST controller scan chain.

Figure 2-21 Controller Setup Chains

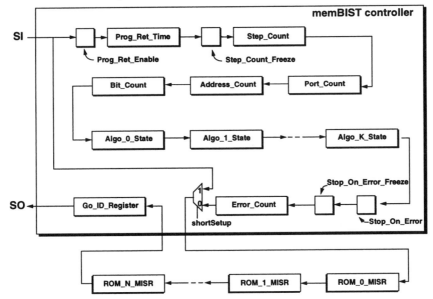

Not all of the registers in Figure 2-21 are necessarily included in each controller. The existence of some registers is dependent on the configuration options chosen when the controller is generated. Table 2-1 on page 54 provides a description of each of the registers in the scan chain.

The LogicVision memory BIST controller supports two modes of scan chain access. In *Short Setup* mode, only the *GO_ID* register and *ROM_N_MISR* registers (if they exist) are placed within the scan chain. This mode is used if only go/no-go test results are of interest. In *Long Setup* mode, the entire scan chain is made accessible.

Choosing either setup mode is accomplished using the controller's *BIST_SETUP[1:0]* pins, as shown in the following table:

BIST_SETUP[1:0]	Selected Mode
0 0	Short Setup
0 1	Long Setup
1 0	Run Default
1 1	Run Normal

Table 2-1 LogicVision Memory BIST Controller Internal Registers

Register Name	Init/ Read	Description
Prog_Ret_Enable	Init	This 1-bit register is used to enable the programmable retention time feature. If a 0 is placed in this register, then the default hardwired retention time is used.
Prog_Ret_Time	Init	This register holds the programmable retention time. If the *Prog_Ret_Enable* register is set to 1, then the value placed in this register overrides the default hardwired retention time information.
Step_Count_Freeze	Init	This 1-bit register is used to force the memory BIST controller to only run through one of its test steps. If a 1 is placed in this register, then only the step number placed in the *Step_Count* register will be executed.
Step_Count	Init / Read	If a 1 is placed in the *Step_Count_Freeze* register, then only the step number placed in this register will be executed by the memory BIST controller.
Port_Count	Read	This register indicates the test port that was being tested when the memory BIST controller stopped. Each test port consists of one logical read and one logical write port or one logical read/write port.
Address_Count	Read	This register indicates the memory address that was being accessed when the memory BIST controller stopped.
Bit_Count	Read	This register indicates the bit of each bit slice that was being tested when the memory BIST controller stopped.
Algo_k_Count	Read	An *Algo_k_Count* register exists for each separate algorithm supported by the configured memory BIST controller. The value stored in one of these registers indicates the algorithm subphase that was being executed when the memory BIST controller stopped.

Register Name	Init/ Read	Description
Stop_On_Error	Init	This 1-bit register is used to enable the stop-on-error feature. A 1 placed in the register forces the memory BIST controller to stop when it encounters n errors, where n is the value stored in the *Error_Count* register.
Stop_On_Error_Freeze	Init	The memory BIST controller places a 1 in this register to stop itself. A 0 must be placed in this register to reenable the controller.
Error_Count	Init	This register is set to the number of errors that must be encountered before the memory BIST controller stops execution.
ROM_n_MISR	Read	A ROM_n_MISR register exists for each ROM tested by the memory BIST controller. Each of these registers contains the signature generated by the MISR used to compress the ROM test results.
Go_ID	Read	This register stores the value of each *Go_ID* signal. A *Go-ID* signal represents the GO/NO-GO status of each individual comparator.

These pins are also used to place the controller in one of two run modes. In the *Run Default* mode, the controller disregards any values scanned into the scan chain registers and uses only hardwired default values. In the *Run Normal* mode, the controller runs based on the values scanned in.

The following six steps are required to extract diagnostic information on the next error.

1. Place the controller in *Long Setup* mode.
2. Scan appropriate values into registers with initialization requirements. In particular, ensure that the error register is provided with the next error number.
3. Place the controller in *Run Normal* mode.
4. Let the controller run until it stops.
5. Place the controller in *Long Setup* mode.
6. Scan out results.

The above steps are repeated for each error until the maximum number of errors is reached or the controller runs to completion. Ideally, when moving from one error to the next, it is desirable to have the controller continue from where it left off. To accomplish this, the scan chain must be initialized with exactly the same values that were just scanned out from the previous error, with the exception of the error counter, which must be set to 1. In the case of DRAMs, it might be necessary to have the controller restart from the beginning for each error, because the time needed to scan values in and out between errors might be longer than the refresh constraints of the DRAM. In this case, the error register must be set with the previous value plus one so that the controller continues until the next error.

Sample Diagnosis of 1-Port SRAM

Assume that a one-port SRAM being tested in the controller's second step is to be diagnosed and that the algorithm being applied is the second of two algorithms supported by the controller. The register values to scan in for each error iteration would be as follows:

Step_Count_Freeze	1
Step_Count	2
Port_Count	Same as value just scanned out (1 in this example)
Address_Count	Same as value just scanned out
Bit_Count	Same as value just scanned out
Algo_0_State	Do not care
Algo_1_State	Same as value just scanned out
Stop_On_Error	1
Stop_On_Error_Freeze	0
Error_Count	1
Go_ID_Register	Do not care

The scanned-out register values of interest for each iteration would be the following:

Port_Count	Indicates which port is failing (always 1 in this example)
Address_Count	Indicates the failing address
Bit_Count	Indicates which bit is failing

References

[1] S.K. Jain and C.E. Stroud, "Built-In Self Testing of Embedded Memories," in *IEEE Design and Test of Computers*, October 1986, pp. 27-37.

[2] B. Nadeau-Dostie, A Silburt and V.K. Agarwal, "A Serial Interfacing Technique for External and Built-In Self-Testing of Embedded Memories," in *IEEE Design and Test of Computers*, vol. 7, no.2 , April 1990, pp. 56-64.

[3] S. Wood, R. Gibson, S. Adham and B. Nadeau-Dostie, "A 5 GHz/s 9-Port Application-Specific SRAM with Built-In Self-Test," in *Proceedings of the 1995 IEEE International Workshop on Memory Technology Design and Testing*, San Jose, August 7-8, 1995, pp. 66-75.

[4] M. Nicolaidis, "An Efficient Built-In Self-Test Scheme for Functional Test of Embedded RAMs," in *Proceedings of the 15th IEEE International Conference on Fault-Tolerant Computing Symposium*, July 1985, pp. 118-123.

[5] M. Nicolaidis, "Transparent BIST for RAMs," in *Proceedings of the IEEE International Test Conference*, 1992, pp. 598-607.

[6] M. Nicolaidis, et. al., "Trade-offs in Scan Path and BIST implementations of RAMs," *Journal of Electronic Testing: Theory and Applications*, vol. 5, no. 2, May 1994, pp. 273-283.

[7] A.J. van de Goor, *Testing Semiconductor Memories: Theory and Practice*. John Wiley & Sons, England, 1991.

[8] P. Mazumder and K. Chakraborty, *Testing and Testable Design of High-Density Random Access Memories*. Kluwer Academic Publishers, Norwell, MA, 1996.

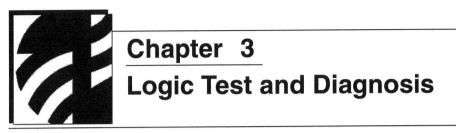

Chapter 3
Logic Test and Diagnosis

This chapter describes more details of the logic test architecture introduced in Chapter 1, "Technology Overview." This chapter describes the circuit preparation and hardware building blocks required to perform an at-speed self-test. In addition, test logic configurations and diagnosis procedures are described. Potential timing issues are also identified and solutions are provided.

Circuit Preparation

A circuit must comply with specific design rules before BIST can be applied to it. This circuit can be the entire UDL of a chip or an embedded core.

As the first step in the Core Design Flow described in Chapter 4, "Embedded Test Design Flow," the rules checker is run to confirm that the design meets requirements.

The design must comply to these design rules only *during test mode*. It is possible to add test-mode logic to achieve conformance to rules (such as a different control path for an asynchronous reset). However, differences between the test-mode circuit and the normal-mode circuit lead to reduced fault coverage.

The tools are intended to work with logic that is *fully synchronous* and *scan-compliant* in test mode.

If a design contains non-scannable blocks, the blocks must be isolated to implement logic BIST for all rule-compliant portions of the design. Hardware collars can be used, as this chapter later describes, to isolate these non-scannable blocks. The collars can be created manually or by using the Legacy Core Test option of the tools to automatically generate the collars.

To test the non-scannable blocks, other test approaches must be used in addition to logic BIST. Add-on options that automatically generate embedded test hardware (including the hardware collars) for many types of non-scannable blocks and merge this hardware into designs are described throughout this book.

Logic Conventions

The following conventions apply to the design rule examples in this chapter:

- All pentagonal connectors are pins (ports) of the top-level design block evaluated by the rules checker. These pins are referred to as *primary inputs* or *primary outputs*.

- The *test-mode (TM) signal* is constant and equal to 1 for the duration of the embedded test.

- All control signals of the logic gates and flip-flops are active on a high level (logic 1) or on the rising edge of the signal.

  For example, a value of 1 applied to the asynchronous reset pin of a flip-flop causes its Q output to be low (0). A 0-to-1 transition on the clock pin of a flip-flop causes the data present at its D input to be transferred to its Q output.

General Requirements

The rules checker verifies that a design meets several general requirements:

● All cells used in the design are modeled with supported primitives or as *black boxes*.

The rules checker supports a subset of Verilog primitives typically used to model combinational cells and a collection of LogicVision primitives required for sequential cells such as flip-flops and latches. The models are to be placed in a directory that is accessible to the rules checker.

Some cells cannot be represented with the primitives available. For such cases, black-box (no, or significantly reduced, functionality) models must be created for those cells.

● All contributing nets in the design are *driven*, meaning that the net is connected to a gate output or the net is a primary input.

● A net does not have multiple drivers unless the drivers each have a tri-state control or a parallel driver's configuration is used.

For each net with more than one driver, the drivers must be *one* of the following:

  • All tri-state gates (pull-up or pull-down cells are also allowed on the net)

  • The same gate type *and* driven by identical signals, as shown in Figure 3-1

● A primary bidirectional pin is driven only by a tri-state gate.

Figure 3-1 Net Driven by Parallel Drivers

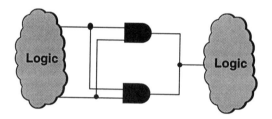

Synchronous Circuits

The circuit under test must comply with synchronous design rules.

The synchronous design rules require a circuit composed of combinational logic and D-type flip-flops. Latches in transparent test mode and retiming latches are also accepted, as discussed in the section "Latches" on page 64. *Combinational logic* is logic that does not contain any flip-flops, latches, or memory components. Figure 3-2 on page 61 shows a conceptual view of a synchronous circuit.

● All flip-flops are clocked from a primary input pin.

● All logic gates form a combinational circuit whose input is driven by primary input pins or flip-flop outputs and whose output consists of primary output pins or inputs to flip-flops.

A *synchronous* design style requires asynchronous set or reset signals to be controlled from primary input pins or to be held at a constant inactive value during test mode.

Figure 3-2 Synchronous Circuit Model

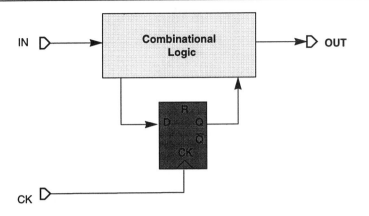

The circuit of Figure 3-2 can contain more than one clock. The circuit complies with design rules even when, during normal operation, the clocks are not in phase or are not operated at the same frequency. The automation tools creates test logic to make the clocks synchronous to each other during test mode.

Clock inputs can be gated using signals generated by the combinational circuit. This feature is very useful for low-power designs. An example is shown in Figure 3-3. Typically, the clock gating circuitry controls the clock of several flip-flops.

Figure 3-3 Synchronous Circuit with Gated Clock

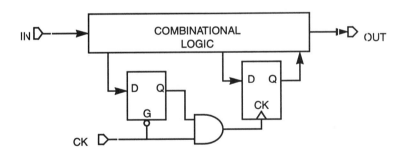

Tri-State Buses

Tri-state buses should be avoided, if possible. In certain cases, however, tri-state buses might be needed to meet the speed or area specifications for a design. The circuit must be designed to ensure that no more than *one* bus driver is on, regardless of the state of the circuit.

Figure 3-4 on page 62 illustrates a fully decoded tri-state bus. In this example, the application logic uses only three bus drivers. A fourth driver is added to ensure that one bus driver is always active, regardless of the state of the circuit. This is to prevent the bus from floating during test mode.

Figure 3-4 Fully Decoded Tri-State Bus

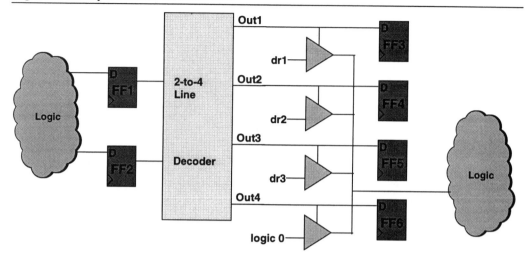

If flip-flops *FF3*, *FF4*, *FF5*, and *FF6* are added to the logic, as Figure 3-4 illustrates, the decoder logic is fully tested using scan or logicBIST. Stuck-at-one faults on the enables of the bus drivers are reported as *potentially detected* faults. A stuck-at-one fault on one of the enables relies on a *bus contention* (two bus drivers are simultaneously active and in opposite states) in order to be detected. In logicBIST, many—if not all—combinations of logic 1 and logic 0 values are applied to *dr1*, *dr2*, and *dr3*. One or more of these combinations of values is likely to create a bus contention that causes a test failure.

Pull-Ups and Pull-Downs

Designers can add pull-up or pull-down cells to propagate faults in the decoding logic and the bus drivers themselves through the enables. In this case, the design must be modified to disable all bus drivers in test mode during scan shift. Also, it must be possible to disable all bus drivers during the capture mode for a subset of the test vectors applied.

The modification to the bus-enable control circuits is not automated and can modify the circuit timing. The design should not allow glitches on the enables of the tri-state bus drivers when switching to capture mode. Otherwise, the test might be invalidated.

If IDDQ test methods are used, additional circuitry is required to disable all pull-up or pull-down cells and to ensure that all tri-state buses are driven (not floating) to eliminate static current.

Figure 3-5 on page 63 shows design modifications that must accompany the addition of a pull-up or pull-down to a tri-state bus.

Figure 3-5 Fully Decoded Tri-State Bus with Pull-Ups

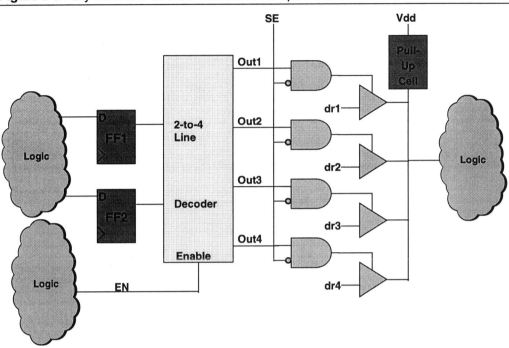

In test mode during scan shift (*SE* = logic 1), decode outputs *Out1* through *Out4* are driven to logic 0, disabling all tri-state bus drivers. This scenario allows sufficient time for the weak pull-up cell to change the voltage on the tri-state bus to approximately *Vdd*.

When *SE* switches to logic 0 (capture mode), the tri-state bus remains disabled if *EN* is equal to logic 0. This circuit allows detection of a stuck-at-one fault on the enable at any of the tri-state bus drivers. For example, if *dr1* is equal to logic 0 and the enable of this driver is stuck-at-one, the tri-state bus is driven to logic 0 versus the expected value of logic 1.

Note that for this circuit, the values of *Out1* through *Out4* must be stable before *SE* switches from logic 1 to logic 0 to avoid glitches on the enables of the tri-state drivers. If *EN* is equal to logic 0 (that is, the tri-state bus remains disabled during capture), a glitch on the enables can cause the tri-state bus to temporarily switch to a logic 0. The weak pull-up cell is typically unable to change the bus to a logic 1 during the capture cycle, leading to a test failure.

Non-Decoded Enables

If the enable signals are not fully decoded during normal operation of the circuit, the circuit illustrated in Figure 3-5 can still be used with a slight modification to achieve a high fault coverage of the logic that normally drives the enables and to reduce the performance impact. The output of the AND gates of the circuit in Figure 3-5 are combined, using an AND gate, with the corresponding functional enable. This way, one driver at most is enabled because the decoder enables only one of the functional enables. In this case, *FF1* and *FF2* are put on a scan chain clocked at a low frequency to reduce the timing criticality of the circuit.

Latches

Because set-reset latches and transparent latches are not fully testable with BIST, this subsection describes techniques for handling the latches during test mode. These techniques can reduce the loss of fault coverage that results from using non-scan memory elements in an otherwise fully scannable circuit. The techniques are *not* automated. The rules checker does not fix violations with respect to untestable latch states. For circuits tested at high speeds, new timing-critical paths through the latches can be introduced. Therefore, it is important to minimize the number of latches that a design uses as state elements.

Set-Reset Latches

Set-reset latches should be replaced with D-type flip-flops whenever possible. The synchronous design rules require a circuit composed of combinational logic and D-type flip-flops. If set-reset latches are used, their latching capability must be disabled during test mode so that they operate like combinational circuits.

In Figure 3-6, an OR gate intercepts the feedback path that the rules checker otherwise reports as an asynchronous loop. An OR gate should be manually implemented in the design to intercept the feedback path. With this approach, the latching capability of the latch is not tested. If the set-reset latch is a cell, the cell must be replaced with individual gates, as shown in Figure 3-6 *(b)*.

The most important benefit is that logic that surrounds the latch can be tested. If the asynchronous loop is not broken during test mode, the rules checker reports a design rule violation.

Figure 3-6 Modifying a Set-Reset Latch During Test Mode

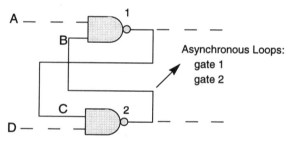

(a) A set-reset latch results in asynchronous loops: A-1-C-2-B and D-2-B-1-C

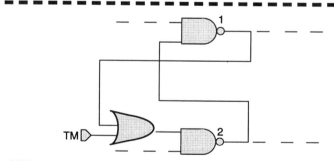

(b) The loops are broken in test mode. However, the latching function is not tested.

Transparent Latches

Transparent latches should be replaced with D-type flip-flops whenever possible. If not, they must be forced into transparent mode when test mode is enabled, as Figure 3-7 illustrates. *Transparent mode* means the gate of the latch is always active; data flushes directly through to the output. The rules checker can then treat the latch as a simple buffer that connects D and Q.

Figure 3-7 Transparent Latch in Test Mode

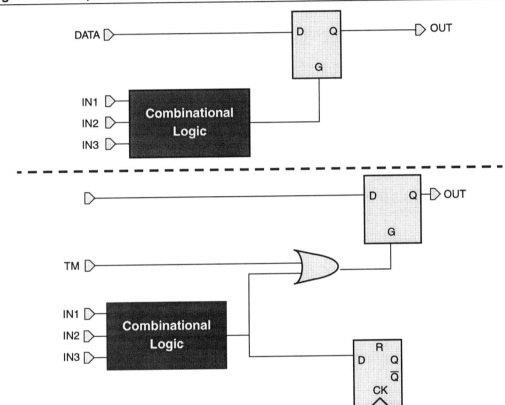

Figure 3-8 on page 66 illustrates how a latch can be used for retiming at the destination clock domain. All flip-flops in the fan-in of the transparent latch must be of the same polarity and must also be of the same polarity with respect to the latch.The source flip-flops do not have to be on the same clock domain. All flip-flops in the fan-out of the transparent latch must be of the same polarity, of the same polarity with respect to the latch, and belong to the same clock domain.

Latches are allowed as retiming elements between clock domains if the gate input is connected to the appropriate phase of the clock signal, as shown in Figure 3-7 and Figure 3-8 on page 66.

When the latch is transparent, the stuck-at-1 fault is not covered on the latch-enable input (G) of the latch, as illustrated in the lower section of Figure 3-7. To detect the faults in the combinational circuit that drives this input in normal mode, a D-type flip-flop can be added as an observation point. The tools replace this D-type flip-flop with a scan flip-flop. The overhead of adding the D-type flip-flop is better justified when the output of the combinational circuit is sufficiently complex.

Figure 3-8 Retiming at the Destination Clock Domain

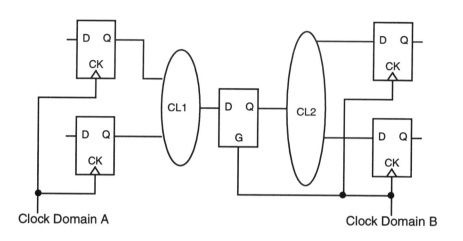

Figure 3-9 Retiming at the Sending End

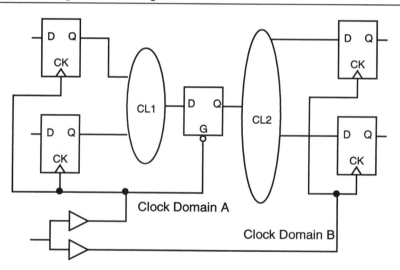

J-K Flip-Flops

J-K flip-flops should be replaced with D-type flip-flops. If not, they should be handled as described in the "Non-Scan Flip-Flops" section.

Non-Scan Flip-Flops

The number of non-scan flip-flops in the design should be minimized; otherwise, fault coverage can be substantially reduced. If a flip-flop is to be excluded from the scan chain, and it is not used as a retiming element, as described in the last section of this chapter, a special scan model must be created for that flip-flop cell and placed in a directory in the search paths that the rules checker uses.

The scan model that is created contains only the module declaration and I/O declarations. This type of scan model is called an *empty* or *black-box* model and is easy to create.

The black-box model causes the flip-flop data inputs to become unobservable. The rules checker ignores all gates in the fan-in tree to the flip-flop until a primary input, a scan flip-flop output, or a *fanout stem* is reached. A fanout stem is a net that is connected to an input of more than one cell.

On the output side, the rules checker treats Q (or QB) as an uncontrollable input to some gates connected to it. If the output of these gates is not observed in test mode, no additional modifications are required.

If the output of these gates is observed in test mode, the uncontrollable input must be intercepted by replacing it with a controllable one or by forcing it to a constant value, as Figure 3-10 illustrates. The latter solution reduces fault coverage but is more straightforward, avoiding the possibility of introducing timing-critical paths.

Figure 3-10 Handling Non-Scan Flip-Flops in Test Mode

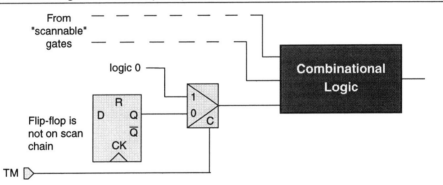

If it is used as a non-scannable element, the same D-type flip-flop cell must have a different name than the name that is used for a scannable element. For non-scannable usage, a new module must be created that consists of a shell wrapped around the D-type flip-flop cell. A black-box model is then created for the new module.

Non-Scannable Blocks

Non-scannable logic blocks in a design can be treated as legacy cores, as described in Chapter 5, "Hierarchical Core Test." Using this method, each block is surrounded with scan flip-flops to observe the logic values arriving at the block's inputs and to control the logic values at its outputs. With this scheme, which provides improved fault coverage, the block must be isolated with multiplexers at all of the block's ports. This isolation is controlled with test-mode signals that drive the select pins of the multiplexers. For small blocks, the scheme can be implemented by manually modifying the RTL or gate-level design.

An alternative is to modify the design to force the outputs from the non-scannable block to known values during logic BIST. In this case, the rules checker removes—or prunes—the untestable logic that surrounds the non-scannable block from the netlist that it creates. The section "Untestable Logic Gates" on page 68 provides more information.

Untestable Logic Gates

The rules checker identifies all gates that are untestable because their outputs cannot change or are not observable because of signals with a static 0 or 1 value that propagate through the circuit. These gates are removed from the design version that the test pattern generation and fault simulation tools use. The *TM* signal from the TAP is used to assert known values on the paths that lead from the outputs of a non-scannable block, as described in the section "Non-Scannable Blocks" on page 67.

Figure 3-11 *(a)*, "Before Pruning," shows an example in which the rules checker removes gates from a circuit because the test-mode signal (TM) is set to a constant value of 1. Gates 1 and 2 are removed because their output values are constant and therefore cannot be tested. Although the output of gate 3 is not constant, the rules checker removes the gate because its output is unobservable.

Figure 3-11 *(b)*, "After Pruning," shows the equivalent circuit created by the rules checker.

Figure 3-11 Example of Untestable Gates

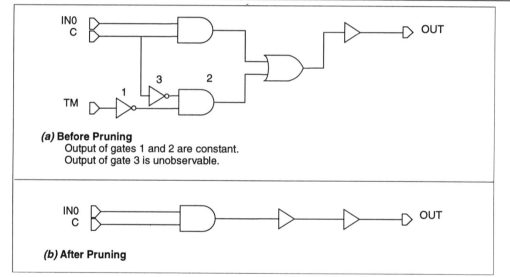

(a) **Before Pruning**
    Output of gates 1 and 2 are constant.
    Output of gate 3 is unobservable.

(b) **After Pruning**

The test generation tool ignores gates that the rules checker prunes. If the number of pruned gates is significant, the circuit should be modified to avoid pruning. Otherwise, functional patterns need to be created by the designer or test engineer to cover the faults associated with the pruned gates.

Logic BIST Building Blocks

Several design structures support the implementation of BIST for the logic. As mentioned in the "At-Speed Logic Test" section on page 19, scan chains and testpoints, as well as a BIST controller, are inserted in the logic to be tested. However, the following other design structures can be added to optimize the test of the logic:

- Test Access Port (TAP)
- Boundary Scan
- Scan-chain router
- Prescaler

The TAP is a standard protocol for accessing all BIST resources and controlling the Boundary Scan (BScan). BScan is the most convenient way to isolate the logic under test from external influence during logic BIST that might affect the expected signature. These design structures are described in Chapter 1, "Technology Overview."

The scan-chain router reconfigures the scan chains during the various icBIST test modes, and the prescaler provides several clock manipulation functions that allow the control of power consumption and the enablement of diagnostics during logic BIST. This section briefly describes these building blocks.

LogicBIST Controller

In summary, icBIST uses a patented, at-speed, multi-frequency, multi-domain logic BIST approach. This approach expands a scan circuit (usually full scan) into a self-test version. The self-test scan circuit uses a pseudo-random pattern generator (PRPG) to generate stimuli and a MISR, based on a cyclic-redundancy checker (CRC), to compress and output results.

Figure 3-12 on page 70 shows a view of the functional blocks in the logicBIST controller.

Figure 3-12 logicBIST Controller Functional Diagram
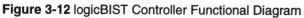

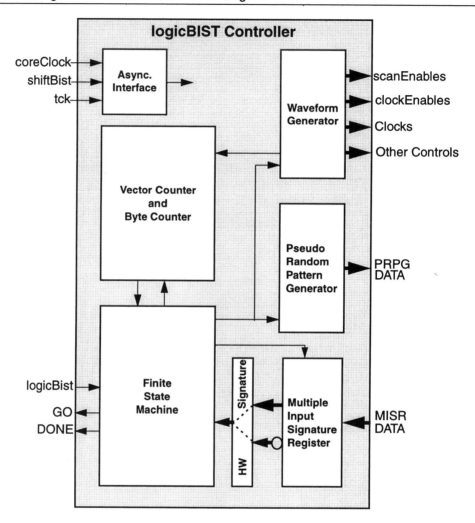

The logicBIST controller consists of the PRPG, the MISR, a state machine that can operate at full application speeds, and the logic required to perform various control functions. These control functions include the following:

- Provide clock signals (*CK*) to flip-flops of clock domains that run at lower frequencies. Two additional frequencies (*coreClock* divided by 2 and by 4) are currently available.

- Provide scan-enable (SE) signals to the scan chains. The *SE* signals are held high while a pseudo-random scan vector is shifted into the scan chains. These signals are then taken low for one clock cycle when the shift operation has been completed to allow the scan flip-flops to capture the circuit responses. There is one scan-enable signal for each frequency. A few options are available to adjust the timing of scan-enable signals (for example, delayed or pipelined).

- Provide clock-enable (CE) signals to flip-flops that have synchronous hold functions. Such flip-flops are used to transmit data into multi-cycle paths. The tools automatically substitute special scan flip-flops, called *priority-hold* flip-flops, when required.

  Within a clock domain in the core block, scan flip-flops that transmit data into multi-cycle paths are grouped into separate scan chains from those occupied by other flip-flops in the domain. The *CE* signals divide the scan shift/capture rate by 2 or 4.

  The *CE* signals are also used with priority-hold flip-flops to adjust timing between the logicBIST controller and the scan chains.

- Count the number of pseudo-random vectors applied and conclude the logic BIST at the proper time. When the logic BIST is invoked using the IEEE 1149.1 TAP during chip, board, or system test, the number of vectors to apply can be specified—from 1 up to the maximum vector count—or the programmed default can be used.

For the default test, the final contents of the signature register are compared within the logicBIST controller block to the hard-wired *golden* signature. In non-default mode, the number of vectors to apply is specified during initialization of the controller, and the signature is shifted out through the TAP to external test hardware, where it is compared to the expected signature. The tools provide the expected signature.

Scan Chains

Typically, multiplexed scan flip-flops are used to replace D-type flip-flops in the scannable logic. Figure 3-13 shows a scan flip-flop.

Figure 3-13 Scannable Flip-Flops

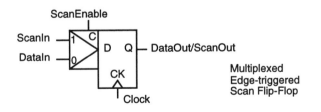

These flip-flops are automatically connected into scan chains. The number of scan chains and the length of the chains can be controlled based on options selected during test development.

Testpoints

Certain types of logic are resistant to test with pseudo-random vectors. An example is a module that contains a 16-bit register that is decoded when all bits are logic 1 values, initiating an operation in a second module. The probability of loading the register during BIST with all logic 1 values is 1 in 65,536. In this example, the second module is not likely to be adequately tested. The output of the decode logic associated with the 16-bit register exhibits a problem with *controllability*.

In other cases, circuit nodes might be difficult to observe. An example is the output of a gate *G1* driving one input of a two-input AND gate for which the other input is almost always forced to zero. The value of *G1*'s output cannot be easily observed at the output of the AND gate. In this case, the output of *G1* exhibits a problem with *observability*.

To increase the fault coverage of logic that is difficult to control or to observe with pseudo-random vectors, testpoints can be inserted. icBIST locates circuit nodes that are difficult to control or to observe and automatically inserts testpoints at the identified nodes.

Testpoints are small blocks of logic that combine a scan flip-flop and zero to two gates. These logic blocks make the problem nodes *controllable* or *observable*. Figure 3-14 illustrates the use of testpoints to improve controllability and observability in a design.

Figure 3-14 Difficult-to-Observe and Difficult-to-Control Nodes

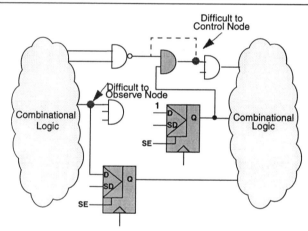

Figure 3-15 on page 73 illustrates the four basic testpoint types:

- Observation point
- AND control point
- OR control point
- XOR control point

Figure 3-15 Supported Testpoints Types (Synchronous Reset)

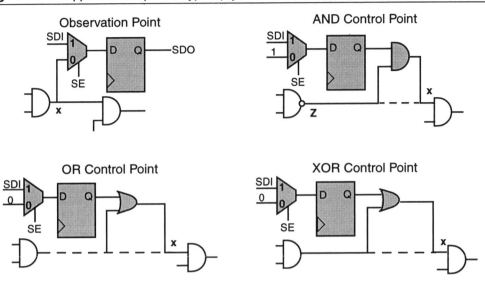

Observation testpoints are used with difficult-to-observe circuit nodes. As the observation point in Figure 3-15 indicates, node *x* is sampled during the scan-capture mode.

● The flip-flop is connected to one of the scan chains of the core subblock to which it belongs.

● The captured value of node *x* is shifted out of the scan chain into the multiple-input signature register (MISR) during scan-shift mode.

In certain cases, it is very difficult to control the value of a circuit node to a logic 1 or 0 value. *Control testpoints* are used to significantly increase the probability that the node is forced to the desired value with a high probability during test mode. In normal operation, the control testpoint must be loaded with its disabling value (0 or 1) so that it does not affect the intended operation of the circuit.

The AND synchronous reset control point enables node *x* to be controlled to a logic 0 value. In test mode, a logic 0 value shifted into the flip-flop causes a logic 0 value to be forced on node *x*.

A synchronous reset control testpoint is disabled by loading the appropriate disable value into its flip-flop. When the TRST pin of the TAP is forced to a logic 0 value, all global scan-enable (SE) signals are forced to a logic 0 value. A single clock cycle is then required to establish the disable value at the Q output of the flip-flop.

For the AND synchronous reset control point, the data input (DI) is connected to a static value of logic 1. With the scan-enable (SE) signal equal to logic 0, the disable value of logic 1 is forced on the Q output of the flip-flop by a single clock.

The testpoint is transparent to the logic except for the propagation delay of the AND gate between nodes *Z* and *x*.

Figure 3-15 also shows the OR and XOR control testpoints. The OR control testpoint enables node *x* to be controlled to a logic 1 value. The XOR control testpoint allows node *x* to be controlled to either a logic 1 or a logic 0 value, depending on the value shifted into the flip-flop.

Asynchronous Reset Control Point

Figure 3-16 shows an asynchronous control point (AND type). Its function during test mode is identical to its synchronous reset counterpart in Figure 3-15 on page 73. However, the asynchronous reset control point uses the *TM* signal to disable its function. When *TM* is driven to a logic 0 value, the testpoint is disabled. This disabling occurs because the output of the NAND gate is forced to 1, making the AND gate that implements the testpoint behave like a buffer. When the *TRST* pin of the TAP is forced to a logic 0 value, the global *TM* signal is asynchronously forced to a logic 0 value.

Figure 3-16 Disabling an Asynchronous Control Testpoint

Asynchronous Control Point (AND)

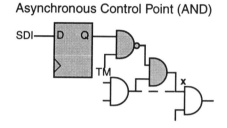

The test generation tool does not check to ensure that the node connected to TM is not stuck-at-one. If the node is stuck-at-one, node *x* is randomly driven to a logic 0, and the application logic is corrupted.

Testpoint Usage

Using synchronous reset testpoints is recommended. Asynchronous reset testpoints contain nodes that are not verified as fault-free. Therefore, unless logic BIST is supplemented with a set of deterministic scan vectors or functional vectors, the fault coverage of a design that uses asynchronous reset testpoints is lower than the same design with synchronous reset testpoints.

The maximum number of testpoints to insert and the fault coverage percentage to achieve for a given number of BIST vectors can be specified to the tools. This feature allows designers to control the trade-off of fault coverage versus test logic overhead and test time for a design. Because testpoints add delay to a logic path, a *critical paths* file can be provided that indicates the nodes on critical paths that cannot have testpoints. No testpoint is added to any of the nodes listed in this file. It is usually possible to reoptimize the circuit with testpoints to restore the original circuit performance. In such a case the area might increase slightly.

Prescaler Block

The prescaler block selects and outputs the clock used by the logicBIST controller. During test, when the logicBIST controller is initialized through the TAP, a clock is selected based on the initialization instructions.

The clock prescaler, which is required for logic BIST, provides these capabilities:

- Selects between three test clock sources (system clock, the output of a PLL connected to system clock, or the IEEE 1149.1 TCK). The *system clock* is the application signal that is designated as the source of the primary test clock during logic BIST.

- Provides scaled versions of the system clock: system-clock frequency divided by two and divided by four. The output of a PLL connected to the system clock can also be scaled such that the PLL frequency can be divided by two or divided by four.

- Stops the test clock (clock freeze) as specified by an instruction from the IEEE 1149.1 TAP. Using this option, it is possible to diagnose potential failures only occurring at high speed during the logic test such that they are not reproducible with a low-speed scan test. In that mode of operation, the scan chains are loaded with a vector and the capture cycle is performed at speed but the high-speed clock is frozen right after the capture operation. The scan chains are then unloaded at low speed, under TAP control, so that a tester can perform logging of the circuit response. No compression is performed by the MISR. Chapter 4, "Embedded Test Design Flow," gives more detail on the diagnostic procedure.

- Reduces the effective scan-shift frequency for lower power consumption during logic BIST. Power reduction is achieved in the prescaler by suppressing a specified percentage of the clocks that is otherwise output from the block. The last few clocks associated with each BIST vector are not suppressed, preserving the quality of the at-speed test. Obviously, test time increases as power dissipation decreases. Once activated, this feature is fully programmable. Test engineers can use the TAP to specify the percentage of clock pulses to be suppressed. Selections are in 1/8 increments: suppress none of 8 pulses, suppress 1 of every 8 pulses, suppress 2 of every 8, and so forth, up to 7 of every 8 pulses.

Scan-Chain Router Block

The scan-chain router configures scan-chain segments in the core block as well as boundary-scan chain segments during three modes of operation:

- **Scan-through-TAP configuration** (logic BIST diagnostic mode and scan mode)

  In the *Scan-through-TAP* configuration, the router uses built-in multiplexers to route the logicBIST controller, the core block's scan chains, and the boundary-scan chains into one serial scan chain.

  This configuration can be used for diagnostic purposes. The diagnostic capability allows for capture of the state of every scan flip-flop in the design at a failing vector and shifts this information out of the TAP's serial data output (TDO) pin.

- **Multi-chain configuration** (parallel scan chains mode)

  The *multi-chain* configuration is similar to the scan-through-TAP configuration in that the configuration includes all scan flip-flops in the design. This configuration can be used to apply test vectors to specific faults that might not be covered in the logic BIST configuration. For the multi-chain configuration, scan flip-flops are segmented into several parallel scan chains that use peripheral pads as scan-in and scan-out ports. Using parallel scan chains for manufacturing tests reduces test time. The number of scan chains is specified during test development. Each ASIC vendor has a different limit for the maximum number of scan chains that can be specified.

- **Logic BIST configuration** (normal logic BIST test mode)

  In the *logic BIST* test configuration, the router connects signals from the logicBIST controller to the appropriate corresponding signals of the core block's scan chains or boundary scan chains. For example, the router connects the following signals:

  - A pseudo-random pattern generator (PRPG) signal to the scan-data input signal of the first flip-flop of each scan chain within the core block

  - A scan-data output signal of the last flip-flop of each scan chain to a multiple-input signature register (MISR) signal

Test engineers select a configuration by issuing a command to the TAP. There is no hard limit on the number of scan chains that can be specified in the logic BIST configuration.

Logic Test Configurations and Diagnosis

This section introduces additional information about the various logic test modes and diagnosis scheme supported by the scan-chain router.

Logic Test Configurations

Three top-level test configurations are available for logic test: scan-through-TAP (used primarily for design debug), multi-chains (used for scan test of the chip), and logic BIST (used for at-speed logic BIST at chip or system test). These top-level logic test modes are in addition to all the test modes supported by the embedded test used in any sub-blocks of the chip.

The designer can choose to implement any or all of these logic test configurations into a chip during the Top-level Flow. The configuration is later selected for use by an operation code loaded into the instruction register of the IEEE 1149.1 TAP controller. Based on the operation code, the TAP controller issues a set of control signals to the scan-chain router to create the requested logic test configuration.

A more complete description of the three logic test configurations is provided below, along with a discussion of related design issues.

Scan-Through-TAP Configuration

The *scan-through-TAP* configuration serially connects all scan flip-flops in logic into a single scan chain. This chain includes scan flip-flops located in the following modules:

- Test-ready logic block
- logicBIST controller
- Scan-chain router
- TAP controller
- Boundary scan register

Figure 3-17 illustrates the scan-through-TAP configuration.

Figure 3-17 Scan-Through-TAP Configuration

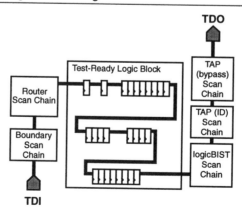

As was briefly mentioned earlier in this chapter, designers typically use the scan-through-TAP configuration for diagnostic testing when debugging a design with scan test or logic BIST failures.

A design issue related to the scan-through-TAP configuration is the potential for hold-time violations during the scan test. The single scan chain that the scan-chain router module creates traverses all clock domains within the chip. In the scan-through-TAP configuration, the test clock for the scan test is a derivative of the TCK signal to the TAP controller. The scan-chain router multiplexes this test clock to the clock tree dedicated to each clock domain in the design. Although an individual clock tree is usually balanced to minimize skew, the skew between trees can be large, creating the potential for data-to-clock hold- time violations during scan test.

To solve this problem, the LogicVision tools allow the implementation of a timing-robust scan test, minimizing the chances of race hazards. To eliminate the potential for hold-time violations during the scan-shift operation, the tools automatically insert negative-edge retiming flip-flops between all concatenated segments.

In addition to supporting scan-chain retiming, the capture-by-domain feature eliminates the potential for hold-time violations during the capture operation. This feature alternates the capture operation between flip-flops that communicate with each other from different clock domains. When the transmitting flip-flop is capturing, the receiving flip-flop is not allowed to enter the capture mode. Likewise, when the receiving flip-flop is capturing, the data entered into the transmitting flip-flop during the last scan-shift operation is held through the capture cycle.

It is important that the designer ensure that the clock distribution within a clock domain is robust during scan. This can be done with the use of static-timing analysis of the scan-through-TAP configuration.

The designer should also note that a multiplexer must be inserted at the root of each clock tree to route the common scan clock to all clock domains. This adds delay to the clock path during the chip's functional mode of operation. The rest of the embedded test structures used for the scan-through-TAP configuration should not impact the functional performance of the chip.

Multi-Chain Configuration

The multi-chain configuration is similar to the scan-through-TAP configuration in that all of the scan flip-flops in the design are included in the configuration.

For the multi-chain configuration, the scan flip-flops are segmented into several parallel scan chains. The number of scan chains to be implemented is specified in an input file read by the LogicVision tools. One of the chains is driven though the TAP where the TDI port is the input to the chain and the TDO port is the output. The remaining specified chains are multiplexed to designer-assigned pad cells of the design. Figure 3-18 on page 78 illustrates the multi-chain configuration.

Figure 3-18 Multi-Chain Configuration

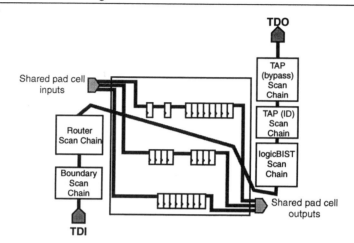

Designers should implement the multi-chain configuration into a design if the test requirements include using scan test for chip manufacturing test. Design issues associated with the multi-chain configuration are identical to those of the scan-through-TAP configuration described on page 76.

Logic BIST Configuration

In the logic BIST configuration, the scan chains in the test-ready logic block and the boundary-scan register are individually linked to PRPG and MISR ports.

During logic BIST, each scan chain captures at a user-defined ratio of the system clock speed, allowing simultaneous test of all clock domains. A clock generated by the logicBIST controller drives each of these domains at one of three available frequencies: main frequency (drives the highest-frequency domains), main frequency/2, and main frequency/4.

Pseudo-random patterns are shifted into the scan chains by the PRPG modules within the logicBIST controller. The output responses are captured and compacted into a signature in the MISR. The timing of the clocks and scan enables needed to achieve high-speed operation is also done by the logicBIST controller.

There are two logic BIST modes available that can be selected by an operation code to the TAP controller when the test is initiated. If the selected test mode is *default*, the logicBIST controller applies a designer-specified number of test vectors to the scan chains and compares the resulting signature to a hard-wired reference signature. Based upon the outcome of the comparison, the controller passes a pass/fail status bit to the TAP controller. If the *normal* test mode is selected, the designer or test engineer can apply any number of pseudo-random test vectors up to the maximum number of available vectors. The designer must specify the maximum number of available vectors to the Top-Level Assembly tool that creates the logicBIST controller. The Analyze tool provides the expected signature for any number of vectors up to the maximum available value.

A *setup* test mode, preceding any *normal* mode, is used to initialize the logic BIST controller. The initialization sequence selects programmable options that the designer builds into the logicBIST controller. When the *normal* test has been completed, the signature contained in the MISR of the logicBIST controller is scanned out through the TAP controller using the scan-through-TAP configuration. The signature is compared to the expected signature stored in the external tester.

Figure 3-19 on page 79 illustrates the logic BIST configuration.

Figure 3-19 Logic BIST Test Configuration

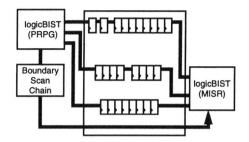

Designers include the logic BIST configuration into chip designs to reduce external tester pattern count, to reduce test time, to perform at-speed logic test, and to provide a test that is reusable at board and system manufacturing.

Design issues associated with the logic BIST configuration are similar to those related to the two scan test configurations. In addition, the designer must be concerned about race hazards associated with the transfer of data between the logicBIST controller and the scan chains.

The designer must assign a clock source from the logicBIST controller to each clock domain within the test-ready logic block. The assignments are made based upon the frequency of each clock domain in functional mode and the anticipated propagation delay through its clock tree. The top-level Assembly tool adds retiming flip-flops and other circuitry as appropriate to provide robust timing between the logicBIST controller and the scan chains of the test-ready logic block.

Logic BIST Diagnostic Scheme

For the logic BIST diagnostic scheme, sets of testbenches are created using the logicbistVerify tool with the runtime option *-diag On*. These testbenches locate the failing BIST vectors through an iterative process. The test engineer specifies a range of BIST vectors to evaluate and the number of subsets to create within the specified range. The logicbistVerify tool then creates a testbench for each subset.

A testbench performs the following tasks for a subset:

- Initializes the logicBIST controller with the proper PRPG, MISR, and vector length values. The TAP operates at TCK rate and the logicBIST controller operates at system-clock rate.

- Instructs the TAP to change to the logic BIST configuration. The TAP reconfigures the scan-chain router to the logic BIST configuration.

- Runs logic BIST at speed until the last vector of the subset is applied and the results are captured in the MISR.

- Extracts the results from the logicBIST controller by scanning out the contents of the MISR and compares these results to the expected value.

- Extracts the contents of all scan flip-flops in the design using the scan-through-TAP configuration. This step is useful only when the failure has been isolated to a specific vector number.

When these testbenches are run on a failing device on the tester, a fail flag (based on the MISR comparison) can be used to identify any testbench with one or more failing vectors.

The test engineer then instructs logicbistVerify to create a new set of testbenches for any segment that contains one or more failing vectors. Using the new set of testbenches with the tester, results are gathered. This process can be continued until the exact failing vector numbers are located.

After an exact failing vector number is determined, the test engineer specifies that vector number as the starting and finishing vector of the range to evaluate. The testbench performs the following tasks:

1. Executes logic BIST for the specified vector.

2. Freezes the test clock.

3. Switches to scan-through-TAP mode.

4. Scans out all flip-flops through the TDO pin using TCK.

However, in this case, the data stream shifted out represents the state of every flip-flop at the failing BIST vector. The logicbistVerify tool provides the expected state of every flip-flop for this vector, allowing the failing flip-flops to be identified. The test engineer can then look at the logic cones associated with the failing flip-flops to further diagnose the problem.

Scan/ATPG Diagnostic Scheme

The second diagnostic mode involves creating a set of deterministic scan vectors to apply while in the scan-through-TAP configuration. The LogicVision signatureAnalyze and scanVerify tools can be used to create a testbench and set of test patterns to support the diagnosis. The procedure for the second diagnostic mode is as follows:

● The test engineer simulates the testbench created by LogicVision's scanVerify tool with the design netlist to confirm that no failures occur when the scan patterns are applied to a defect-free circuit.

● The test engineer converts the WGL or SVF-formatted test pattern set from LogicVision's scanVerify tool to a test vector set in the format required by the ATE hardware. To perform this conversion, a company's proprietary translator program, or a commercial translator, can be used.

● The engineer tests the devices to diagnose using the converted test vector set and the ATE hardware. The engineer captures the serial data shifted out of the TDO pin of the device and compares the captured data to the expected result. This approach allows the failures to be traced to the individual flip-flops in the scan chains.

Timing Issues and Solutions During Logic Test

Several issues must be addressed to avoid hold and setup time violations during logic test. These issues arise because testing is performed at application speeds and all logic is tested simultaneously, in a synchronous manner, on designs that have multiple clock domains. Fortunately, the timing issues can be addressed with robust solutions. This section covers these problems and their corresponding solutions and describes the clock sources available from the logicBIST controller.

The most likely timing problems are in the following areas of the logic architecture:

● Interface during the scan shift operation between the logicBIST controller and the scan chains.

● Distribution of the scan-enable signal to the scan-enable ports of large numbers of multiplexed scan flip-flops, during one at-speed clock period as required by the scan test methodology

● Interface during the capture operation between scan chains that belong to different clock domains

● Multi-cycle paths

logicBIST Controller Clock Sources

Figure 3-20 illustrates the clock sources that derive from the logicBIST controller.

● As described in the "Prescaler Block" section on page 74, the output from the clock prescaler is the source for the master test clock (*MCLK*) of the logicBIST controller.

● Buffer symbols labeled *M*, *A*, *B*, *C*, and *D* in Figure 3-20 represent multiple clock distribution networks.

● Test clock sources are multiplexed with functional clocks before the clock distribution networks. Figure 3-20 does not show the multiplexers.

A design might not require all available logicBIST clock sources. The number of clock distribution networks driven by a given clock source from the logicBIST controller is design dependent. A design with multiple clock domains—operating at the same frequency—might have a large number of clock distribution networks connected to one clock source.

The logicBIST controller is always driven from the clock distribution network labeled *MCLK* that is sourced by the signal *1E* in Figure 3-20. Any one of the signals listed below, which Figure 3-20 illustrates, can source the clock distribution networks for all other clock domains.

● **1E**—output of the prescaler

● **1**—MCLK (master test clock)

● **2**—clock [2] output port from the logicBIST controller

● **4**—clock [3] output port from the logicBIST controller

Figure 3-20 logicBIST Controller Clock Sources

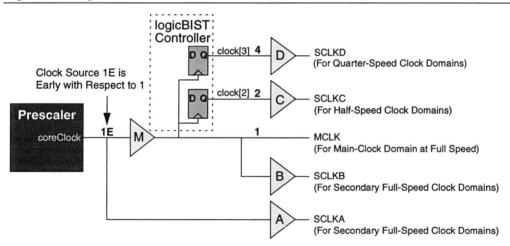

In Figure 3-20, domain *SCLKD* runs at one quarter of the *MCLK* frequency and is connected to *clock[3]*. Domain *SCLKC* runs at one half the *MCLK* frequency and is connected to *clock[2]*. Both *SCLKA* and *SCLKB* run at full frequency. *SCLKA* transitions earlier than *SCLKB* because *SCLKA* is connected to the output of the prescaler and *SCLKB* is driven from the *MCLK* clock distribution tree.

Interface for logicBIST Controller and Scan Chains

Figure 3-21 shows the interface during the scan-shift operation between the logicBIST controller and the scan chains.

Figure 3-21 logicBIST Controller Timing Interface for Scan Chains

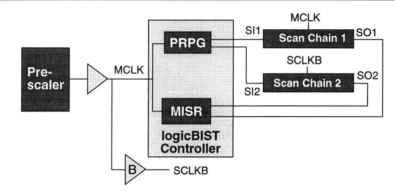

In Figure 3-21, *Scan Chain 1* belongs to the clock domain that is driven by the *MCLK* clock distribution network. The logicBIST controller is always driven by *MCLK*. *Scan Chain 2* belongs to a different clock domain that is driven by *SCLKB*. As shown in the figure, *SCLKB* is delayed from *MCLK* by the propagation delay *tpdB* of the clock distribution network B. It can be assumed that, in test mode, the frequency of *MCLK* is identical to the frequency of *SCLKB*.

Figure 3-22 illustrates the timing relationship between PRPG signals, clocks to the scan chains, and signals to the MISR in Figure 3-21. For the technology in this example, the minimum data-clock hold-time specification for the flip-flops used in the scan chains is 0 nanoseconds.

Interface PRPG to Scan Chains

In this example, a hold-time violation occurs for the first flip-flop in *Scan Chain 2* in Figure 3-21.

- The data *SI2* to the first flip flop changes *before* the occurrence of the clock *SCLKB,* as shown by the *th1* parameter in Figure 3-22 on page 83. The correct relationship of the signals is for *SI2* to change *after* the active edge of *SCLKB*, which satisfies the data-clock hold-time specification.

- Because the data *SI1* to the first flip-flop in *Scan Chain 1* changes after the occurrence of *MCLK,* the minimum hold-time specification is met for this flip flop.

- The data *SO2* from the last flip-flop in *Scan Chain 2* changes *after* the active edge of *MCLK*. Therefore, the hold-time specification is met for the flip-flop in the MISR of the logicBIST controller.

Because *MCLK* drives both the logicBIST controller and *Scan Chain 1*, timing violations do not occur provided that the clock distribution network *M* is designed properly. The network should be designed so that clock skew is less than the minimum propagation delay of any flip-flop driven by the network.

Figure 3-22 logicBIST Controller Timing Waveform for Scan Chains

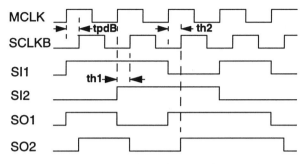

th1 is the data hold time (SI2 – SCLKB) of the first flip-flop in Scan Chain 2.
th2 is the data hold time (SO2 – MCLK) of the flip-flop in the MISR.

Sufficient delay between the PRPG output and *SCLKB* can be automatically inserted, eliminating the hold-time violation.

In the example, logicBIST adds delay to the path between the PRPG output and the input to *Scan Chain 2*, as Figure 3-23 illustrates. The delay results from using a retiming flip-flop that is clocked on the negative edge of *MCLK*.

Figure 3-23 Delayed PRPG Signal to Correct Hold-Time Violation

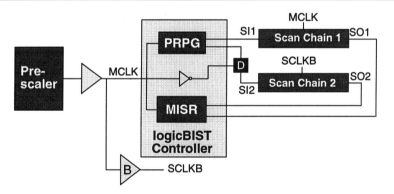

Note that in Figure 3-24 on page 84 *SI2* now occurs after *SCLKB* and *th1* meets the flip-flop data-to-clock hold-time requirement.

Figure 3-24 Timing After Inserting Delay into SI2 Path

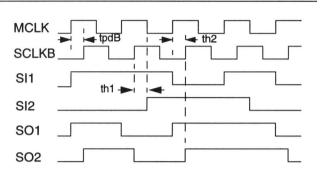

th1 is the data hold time (SI2 – SCLKB) of the first flip-flop in Scan Chain 2.
th2 is the data hold time (SO2 – MCLK) of the flip-flop in the MISR.

The solution above works well *except* at high frequencies in which the delay *tpdB* of the clock distribution network is greater than one half the clock period of *MCLK* and *SCLKB*.

In this case, retiming from the negative edge of *MCLK* does not delay *SI2* sufficiently to satisfy the
flip-flop hold-time requirement, as Figure 3-25 illustrates. *SI2* transitions before the occurrence of
SCLKB, producing a value for *th1* that violates the minimum hold-time specification.

Figure 3-25 Timing Violation: *tpdB* > Half of *MCLK* Period

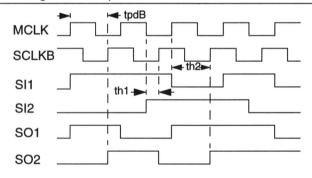

th1 is the data hold time (SI2 - SCLKB) of the first flip-flop in Scan Chain 2.
th2 is the data hold time (SO2 - MCLK) of the flip-flop in the MISR.

The high-frequency problem can be fixed by specifying to logicBIST that the clock of the second
scan chain is driven from the clock source *1E*, as shown by *SCLKA* in Figure 3-20 on page 81.

Figure 3-26 on page 85 illustrates that these actions eliminate the *th1* hold-time violation:

● Assignment of the *1E* clock source (*SCLKA*)

● Insertion of a negative-edge retiming flip-flop into the path from the PRPG to *SI2*

The active edges of *MCLK* and *SCLKA* are now separated by the difference in the propagation
delays *tpdM* and *tpdB* of their respective clock distribution networks.

Figure 3-26 Using Clock Source 1E with Negative-Edge Retiming

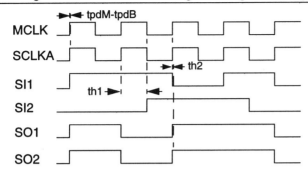

th1 is the data hold time (SI2 - SCLKA) of the first flip-flop in Scan Chain 2.
th2 is the data hold time (SO2 - MCLK) of the flip-flop in the MISR.

Interface Scan Chains to MISR

Now that the high-frequency problem is solved by using the clock source 1E with a negative-edge retiming flip-flip, as described on page 84, *th2* is an issue. The relationship between transitions on *SO2* and *MCLK* depends upon the exact relationship between the delays of the clock distribution networks *tpdM* and *tpdB*. If *tpdM* is greater than *tpdB*, a hold-time violation results at the flip-flop in the MISR that is driven by *SO2*.

The hold-time violation between *Scan Chain 2* and the MISR can be easily corrected. Delay can be automatically inserted in the return path from *SO2* of *Scan Chain 2* to the MISR.

In this case, a negative-edge retiming flip-flop is inserted into this path, in addition to the retiming flip-flop in the path from the PRPG to *SI2*. Both retiming flip-flops are clocked by the negative edge of *MCLK*. Figure 3-27 shows these changes.

Figure 3-27 Delaying Signals to and from Scan Chain 2

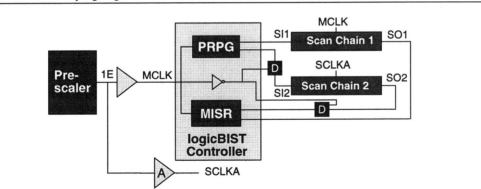

Figure 3-28 on page 86 illustrates the resulting waveforms from delaying the signals to and from *Scan Chain 2*.

Figure 3-28 Using Clock Source 1E with Negative-Edge Retiming

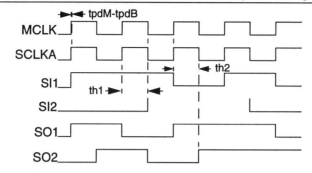

th1 is the data hold time (SI2 - SCLKA) of the first flip-flop in Scan Chain ;
th2 is the data hold time (SO2 - MCLK) of the flip-flop in the MISR.

Potential timing problems in clock domains 2 and 4 can be addressed as described above provided that the propagation delay through their clock distribution tree is less than one half of the clock period of *MCLK*. If the delay is more, strobing techniques are used to transfer scan data to and from the scan chains clocked at lower frequencies. Strobing flops, clocked by *MCLK,* safely sample PRPG output data and hold the scan data for two or four clock cycles before it is transmitted to the actual scan chain through an optional retiming flip-flop clocked by *SCLKC* or *SCLKD*. Similarly, the output of the scan chains connected to the MISR can be strobed at the appropriate time to guarantee a repeatable signature. The strobing mechanisms are inserted automatically based on user input.

Scan-Enable Signal Distribution

Timing problems associated with the scan-enable signal distribution network for scan chains can occur. The scan-enable signal must be propagated to the scan-enable ports of all multiplexed scan flip-flops in a clock domain during one clock period without violating flip-flop setup and hold-time specifications. Figure 3-29 illustrates these scan-enable timing requirements.

Note that the skew between the arrival times of the scan-enable signals at the various flip-flops is not important.

Figure 3-29 Scan-Enable Timing Requirements

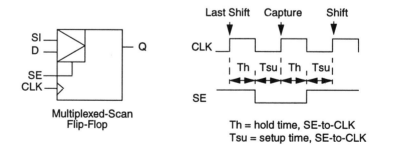

To avoid the possibility of violating the hold-time specifications for any of the multiplexed-scan flip-flops, the scan-enable (SE) signals can be delayed using the same approach described in the "log-icBIST Controller Timing Interface for Scan Chains" section on page 82.

The transitions of the scan-enable signal must reach each multiplexed scan flip-flop in time to meet the setup time specification of the flip-flop. There are two suggested methods for distributing scan-enable signals:

- The first distribution method entails using a clock distribution network supplied by the ASIC vendors as if the scan controls were clocks. The advantage of this technique is that it is trivial to implement for the ASIC designer. The disadvantage is that, for designs that have multiple clock domains, the approach requires a large number of signal distribution networks. Typically, two distribution networks per domain are required: one for the scan-enable and one for the clock.

  When the signal distribution networks are implemented as clock spines, the number of networks is normally limited to a relatively small number (5 to 10). This small number is due to the impact of the spines on routability of a design during physical layout. When these signal distribution networks are implemented using custom synthesized buffer trees, the requirement for a large number of networks is not a problem.

  The propagation delay through a signal distribution network must be small for high frequency designs. For large distribution networks, the propagation delay can easily be 4 to 5 nanoseconds. At frequencies less than 100 MHz, this method is a workable solution.

- The second distribution method entails routing the enable signals as normal data signals and using pipelining stages to distribute the delay across several clock cycles. The signals that are needed to use pipelining can be automatically created. The advantage of this technique is that it does not require the signal distribution networks described in the first method. It also works for high frequencies because it does not require the enable signal to reach the flip-flops in a single clock period. The disadvantage is that the pipelining flip-flops must be manually inserted and distributed throughout the design. These pipelining flip-flops must also be bypassed during normal low-speed scan operation because the IEEE 1149.1 protocol does not support pipelined scan-enable signals.

For ASIC families that support clock-tree synthesis, it is efficient to use pipelining flip-flops and the signal distribution trees together. Each region of a domain has a local pipelining flip-flop but the distribution from this local pipelining flip-flop, and all scan flip-flops in the region, occurs using a signal distribution tree. The propagation delays of the signal distribution trees can be controlled to acceptable values using this approach. The delay of a signal distribution tree is a function of the number of loads that must be driven by the tree. If the number of loads is reduced, the propagation delay can be reduced accordingly.

For example, assume a high-frequency domain with 10000 flip-flops. Distributing the scan-enable signal to all flip-flops within the clock period might be impossible to accomplish, whereas distributing SE to 500 flip-flops *is* possible. To achieve such distribution, the domain is separated into twenty regions using local pipelining flip-flops per region.

Figure 3-30 on page 88 shows a domain separated into four regions, each using one stage of pipelining flip-flops. A synthesized buffer tree can be used to distribute *SE'*.

Figure 3-30 Example of a One-Stage, Four-Region Pipelining Scheme

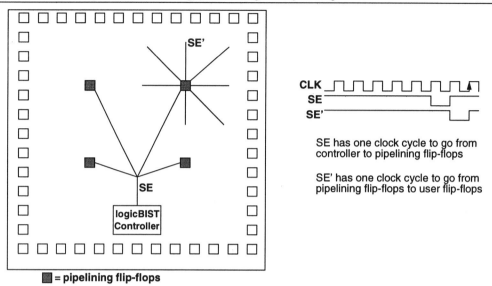

CLK

SE

SE'

SE has one clock cycle to go from
controller to pipelining flip-flops

SE' has one clock cycle to go from
pipelining flip-flops to user flip-flops

■ = pipelining flip-flops

Interfaces for Flip-Flops in Different Clock Domains

Action must be taken to avoid timing violations that result from the interface during the capture
operation between scan chains in different clock domains. The potential problem is shown in
Figure 3-31 and Figure 3-32 on page 89. In Figure 3-32, the application logic is configured so that
FFA in *Domain A* is connected to *FFB* in *Domain B*. In the application mode, this interface can be
controlled in various ways to prevent timing violations.

Figure 3-31 Interdomain Communication During Capture Cycle

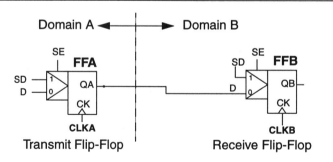

In test mode, scan test or logic BIST exercises all testable logic simultaneously. Hold-time violations
can occur in test mode, as shown in Figure 3-32 on page 89. Such violations require design
considerations.

Figure 3-32 Hold-Time Violation During Capture Cycle

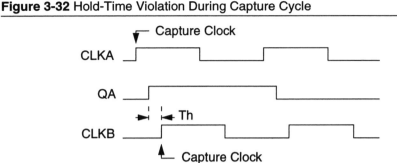

In this example, *QA* of *FFA* changes to a 1 as a result of the capture clock. *FFB* might capture the 1 instead of the correct capture value of 0. The transition of *QA* must occur after *CLKB* by the amount of the hold- time specification of *FFB*.

Before making any scan flip-flop substitutions, all clock domains within the netlist can be automatically identified. The scan insertion tool locates each flip-flop that transmits data to a clock domain that is different from its own domain. It also locates each flip-flop that receives data from a clock domain that is different from its own.

When replacing a non-scan flip-flop with a scan version, this tool uses a standard multiplexed scan flip-flop in cases where the flip-flop does not transmit data to or receive data from a clock domain different from its own.

In cases where the flip-flop is a *transmit* or *receive* flip-flop, a special flip-flop is substituted. The scan flip-flops substituted for *transmit* and *receive* flip-flops are shown in Figure 3-33. In the figure, the transmit flip-flop is a *priority data* flip-flop.

Figure 3-33 Transmit and Receive Flip-Flops

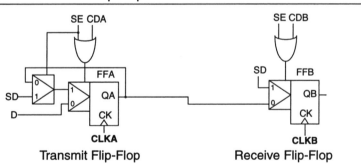

Transmit Flip-Flop   Receive Flip-Flop

Using these flip-flops with the capture-disable signals from the scan-chain router block avoids hold-time violations during a scan test or logic BIST capture cycle. Examples of capture-disable signals are *CDA* and *CDB,* as shown in Figure 3-33. The capture-disable signals are controlled so that they are never both at a logic 0 during the same capture cycle.

For the example of Figure 3-34 on page 90, during a capture cycle the scan-enable signal SE is held at a logic 0. If *CDA* is at a logic 1 and *CDB* is at a logic 0, the capture of the transmit flip-flop is suppressed.

The transmit flip-flop holds its previous state while the receive flip-flop is allowed to capture. Because the transmit flip-flop cannot change state during the capture cycle, there is no potential for a hold-time violation at the receive flip-flop.

Figure 3-34 Capture by the Receive Flip-Flop

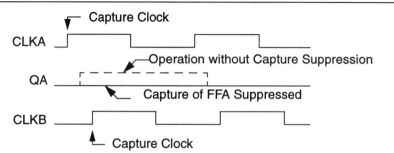

If *CDA* is at logic 0 and *CDB* is at logic 1, the transmit flip-flop is allowed to capture but the receive flip-flop is left in shift mode; hence, its capture is suppressed. Because the receive flop is not sampling the signal from the transmit flip-flop, there is no potential for a hold-time violation.

Figure 3-35 illustrates the waveforms associated with capturing by the transmit flip-flop.

Figure 3-35 Capture by the Transmit Flip-Flop

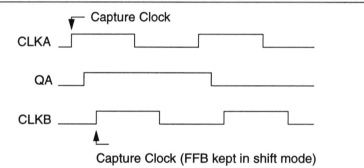

The tool automatically defines the capture-disable signals needed from the scan-chain router. Based on its knowledge of the inter-domain communication, the tool creates the necessary number of *capture groups* and connects the capture-disable (*CD*) pin of each transmit and receive flip-flop to a capture-disable net that corresponds to one of these capture groups. A scan-chain router that generates the capture-disable signals and connects these signals to the appropriate capture-disable nets is automatically generated.

Interfaces to Multi-Cycle Paths

A *multi-cycle path* is a path that requires more than one clock cycle for the data to traverse the path. In the design of certain functionality types, it might be necessary to incorporate multi-cycle paths into a design. For example, synchronous counters that use multi-cycle paths can run at a higher frequency versus counters constructed with the same technology but without multi-cycle paths.

During at-speed test with logicBIST, in the clock period between the last shift clock and the capture clock, all data is expected to traverse all logic paths and meet flip-flop setup time requirements to the capture clock. Obviously, this requirement cannot be fulfilled for multi-cycle paths.

The flip-flops in a design that transmit data into multi-cycle paths must be identified in an input file. The scan insertion tool substitutes special multiplexed scan flip-flops for these components. The flip-flops are controlled by a clock-enable signal, *CE*, generated by the logicBIST controller. Figure 3-36 shows these flip-flops.

Figure 3-36 Multi-Cycle Path Scan Flip-Flops

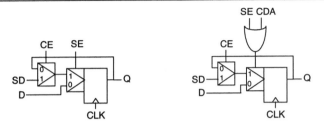

For cases in which the flip-flop acts as a source to a multi-cycle path within its own clock domain, during shift mode (a scan vector with pseudo-random data is loaded), the CE signal controls the shift rate. During a clock cycle where SE is held at logic 1 and CE at logic 0, the previous scan data is held in the flip-flop (shift is inhibited). Shift rates of one half or one quarter of the rate of the master clock (*MCLK*) can be implemented. By reducing the shift rate, the time from the last shift clock to the capture clock increases, allowing enough time for multi-cycle paths to settle.

If the flip-flop sources data to a multi-cycle path in a different clock domain from its own domain, then the second flip-flop type is substituted. Capture is suppressed when the *CDA* signal is held at logic 1. The hold-time margin that is provided around the capture clock corresponds to the shift rate of the multi-cycle path source. Therefore, it is two clock cycles of hold time if the shift rate is half the maximum frequency and four clock cycles if the shift rate is one quarter of the maximum frequency.

References

[1] B.H. Seiss, P. Trouborst and M. Schulz, "Test Point Insertion for Scan-Based BIST," in *Proceedings of the 1991 IEEE European Test Conference*, pp. 253-262.

[2] R. Lisanke, F. Brglez, A. J. Degeus and D. Gregory, "Testability Driven Random Test-Pattern Generation," *IEEE Transactions on Computer-Aided Design*, Vol. CAD-6, November 1987, pp. 1082-1087.

[3] B. Nadeau-Dostie, D. Burek and Abu S.M. Hassan, "ScanBist: A Multifrequency Scan-Based BIST Method," *IEEE Design & Test of Computers*, Vol. 11, No. 1, Spring 1994, pp. 7-17.

[4] P.H. Bardell, J. Savir and W.H. McAnney, *Built-In Test for VLSI: Pseudorandom Techniques*. J. Wiley and Sons, March 1988.

[5] P.H. Bardell and W.H. McAnney, "Parallel Pseudo-Random Sequences for Built-In Test," in *Proceedings of the 1984 International Test Conference*, pp. 302-308.

[6] E.B. Eichelberger, T.W. Williams, "A Logic Design Structure For LSI Testability," *Journal of Design Automation and Fault-Tolerant Computing*, Vol. 2, No. 2, May 1978, pp. pp. 165-178.

[7] B. Koenemann, J. Mucha and G. Zwiehoff, "Built-In Logic Block Observation Techniques," in *Digest of Papers of the 1979 Test Conference*, pp. 37-41.

[8] J.P. Hayes and A.D. Friedman, "Test Point Placement to Simplify Fault Detection," in *Proceedings of the 1974 Fault Tolerant Symposium*, pp. 73-78.

[9] M. Abramovici, M.A. Breuer and A.D. Friedman, *Digital Systems Testing and Testable Design*. IEEE Press Marketing, 1994.

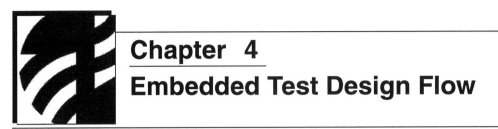

Chapter 4
Embedded Test Design Flow

This chapter discusses the design flow for adding embedded test into a design. The LogicVision process and tool flow are used as examples throughout the chapter.

There are three phases to the process of adding embedded test to a design.

1. Add embedded test to any sub-blocks in the top-level logic block.

    In this phase, sub-blocks such as memories should have BIST added. Sub-blocks such as legacy cores need access and isolation added.

2. Prepare the top-level logic block for scan test or logic BIST.

    In this phase, the sub-block's embedded test needs to be made accessible, and the logic needs to be rule checked and made scannable.

3. Add embedded test structures to the top-level block of the chip.

    In this phase, top-level embedded test structures, such as the logicBIST controller, need to be added. All embedded test features in the chip need to be made accessible from the pins of the chip, usually through a TAP. Finally, the test patterns for use on the manufacturing ATE need to be output.

By adding embedded test to the sub-blocks before preparing the logic blocks for scan or logic BIST, the logic used in the embedded test structures of the sub-blocks can be made ready for scan or logic BIST testing at the same time as the rest of the logic. Also, the tools that operate on the sub-blocks can output models or information about the sub-blocks and their embedded test structures for use by the tools in the subsequent phases.

In each of the above three phases, various LogicVision tools perform the following tasks:

1. Generate the embedded test structures and access mechanisms.

2. Assemble the embedded test structures into the design.

3. Analyze a design with or without the embedded test structures.

4. Verify a design that contains embedded test by using testbenches.

The design flow and automation tools are compliant with a synthesis and simulation-based design flow using standard hardware description languages (Verilog and VHDL). Industry-leading synthesis, simulation, and verification tools are supported.

The Generate and Assemble tools work at the register-transfer level (RTL), with the exception of the Generate tool for scan and testpoint insertion, which requires a netlist. The Analyze tool generally works at the gate level, because they require the structural netlist in order to precisely determine measures such as coverage or to compute signatures and test vectors. The Verify tool can be used at both the RTL and gate levels during RTL simulation or gate-level timing verification.

Technology libraries for use with the LogicVision tools are generally available from semiconductor vendors or directly from LogicVision. In some cases, designers might need to create their own libraries if they use special cells or proprietary, in-house technologies.

Although there are many steps in the process, the flow is highly automated and well integrated. An experienced designer can complete the LogicVision design flow in less time than when using scan/ ATPG and ad-hoc testability. Using the flow, it is possible to obtain very high, at-speed coverage while avoiding manually writing numerous test patterns.

Overview of the Design Flow

The LogicVision design flow for embedded test encompasses three phases:

● The Sub-Block Flow—prepares sub-blocks for embedded test

● The Prepare Logic Flow—prepares top-level logic blocks for embedded test

● The Top-Level Flow—completes the embedded test capabilities for the chip

Figure 4-1 illustrates the design architectural view.

Figure 4-1 Architectural View of the LogicVision Design Flow

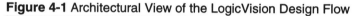

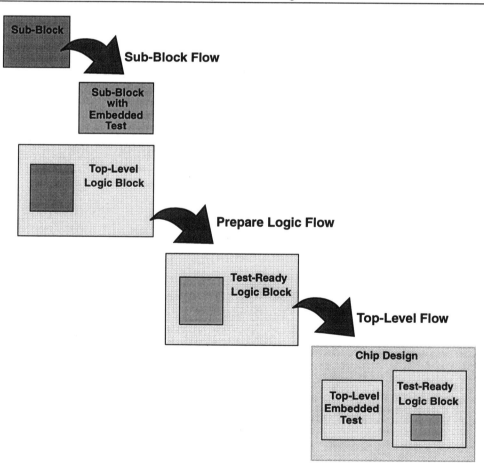

Adding Embedded Test to Sub-Blocks

The purpose of this phase of the LogicVision design flow is to add embedded test to any sub-blocks in the top-level logic block. For sub-blocks such as memories, BIST must be added. For sub-blocks such as legacy cores, *access* and *isolation* must be added. After the embedded test structures are added, they must be synthesized and verified to confirm correct operation. Figure 4-2 illustrates the sub-flow for this phase.

Figure 4-2 The Sub-Block Flow

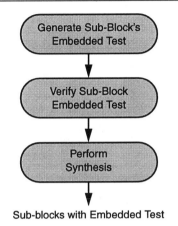

Sub-blocks with Embedded Test

Preparing the Top-Level Logic Block

The purpose of this phase of the LogicVision design flow is to prepare the top-level logic block for scan test or logic BIST. During this phase, the sub-block's embedded test is made accessible and the logic is rule-checked and made scannable. After the top-level logic block is ready for test, it can be integrated into a new version of the top of the chip design that incorporates embedded test. Figure 4-3 on page 96 illustrates the sub-flow for this phase.

Figure 4-3 The Prepare Logic Flow

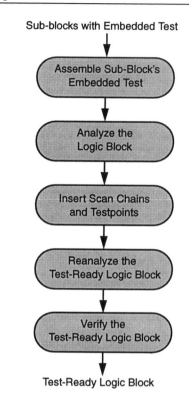

Sub-blocks with Embedded Test

Assemble Sub-Block's
Embedded Test

Analyze the
Logic Block

Insert Scan Chains
and Testpoints

Reanalyze the
Test-Ready Logic Block

Verify the
Test-Ready Logic Block

Test-Ready Logic Block

Adding Embedded Test to the Top Level of the Chip

The purpose of this phase of the LogicVision design flow is to add embedded test structures, such as the logicBIST controller, to the top-level block of the chip. Then the chip-level architecture is modified to provide access from the chip pins to the embedded test structures. Typically, a TAP is added. The TAP only requires four (or five when including the optional reset pin) package pins and serves as the embedded test manager. The Verify tool creates testbenches for verification of the chip-level embedded test architecture. After verification is complete, the designer creates test patterns for use on the manufacturing ATE.

After this phase is completed, the embedded test version of the chip is finished and all test patterns are ready for manufacturing.

Figure 4-4 illustrates the sub-flow for this phase.

Figure 4-4 The Top-Level Flow

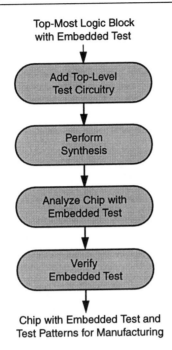

Overview of the Tool Set

The LogicVision tools are targeted for an HDL-based synthesis and verification design flow. The tools, classified into the general categories of *Generate, Assemble*, *Analyze*, and *Verify*, perform the following tasks:

1. Create embedded test structures using the Generate tool. The Generate tool implements the following embedded test capabilities:

    - Scan chains and testpoints
    - IEEE 1149.1 TAP and Boundary Scan
    - Logic BIST
    - Embedded memory BIST
    - Legacy core access and isolation
    - Hierarchical core-based logic test
    - Phase-locked loop BIST
    - External memory BIST
    - Board-level, at-speed interconnect test
    - Fault insertion

2. Insert the embedded test structures into the design and route their connections to the appropriate control structures or chip pins using the Assemble tool.

3. Check rules, analyze the testability, and generate test vectors or signatures for a design, before or after embedded test insertion, using the Analyze tool.

4. Create testbenches to verify that the embedded test structures correctly operate in the design and output the manufacturing test patterns using the Verify tool.

Generating Embedded Test

The Generate tool creates synthesizable Verilog or VHDL RTL descriptions of embedded testers and test infrastructure (scan chains, controllers, test collars, boundary-scan cells, clock prescaler, scan-chain router, and test access port). Unlike the other Generate tool, the Scan Generate tool inserts scan chains and testpoints into the gate-level netlist.

Each of the Generate tools, except the Scan Generate tool, creates synthesis scripts that accompany the RTL. These scripts are used to create technology-specific, gate-level descriptions of the embedded test structures. These scripts can be modified to add design-specific constraints.

A helper tool for the Generate tools, called the *Prep* tool, can be used to assist in transforming certain design-related information into data used by the Generate tools and, later, by the Assemble tool.

Assembling Embedded Test into the Design

The Assemble tool automates the process of incorporating and connecting the embedded test structures created by the Generate tools into a design. The Assemble tool works on either an RTL or gate-level design. The tool instantiates the embedded test structures into the design and route all connections between the embedded test structures and the functional modules and pins in the chip. The Assemble tool works in a completely hierarchical manner, instantiating test structures and making connections across multiple levels of the design hierarchy.

Analyzing the Design with Embedded Test

The Analyze tool helps the designer improve the testability of the logic in the design and create the test patterns and signatures for logic test. This tool performs the following tasks:

● Extracting the network of embedded test structures that have been inserted into the design and checking the extracted information against connection rules.

● Verifying that logic testability rules are satisfactorily met.

● Fault-simulating the logic blocks to determine the coverage.

● Locating nodes in the logic that are resistant to test with logic BIST and modeling the improvement in fault coverage that results from the use of testpoints at these nodes.

● Running LogicVision's ATPG tool to create deterministic test vectors for scan testing the logic.

● Computing signatures for logic blocks tested with logic BIST.

Verifying the Design with Embedded Test

The Verify tool creates testbenches to verify the proper operation of the embedded testers and test the design infrastructure.

The testbenches initialize and operate the various test modes of the embedded test structures, then return test completion values (DONE flags, pass/fail signals) for observation. They are used as input to simulation tools. The testbenches also provide diagnostic information that is useful for debugging.

The Verify tool also outputs test patterns for using the embedded test on the manufacturing ATE. Patterns can be output in WGL for chip testing and SVF for board-level testing.

Required Libraries and Setup

Libraries are required for several of the tools used in the flow. These libraries are generally technology-dependent and describe characteristics of cells in a form that is understandable by the tools. These libraries can be created by the designer on a per-design basis, although generally they can be provided either by the chip manufacturer or by LogicVision.

LogicVision libraries are required to support the following operations:

● Memory library
● Scan and testpoint insertion
● Pad cell insertion

Adding Embedded Test to a Sub-Block

The first phase of the LogicVision design flow is to add embedded test structures to any sub-blocks that need their own embedded tests. Typical candidate sub-blocks are any non-scannable modules in the design, such as memories, legacy cores, PLLs, and ADCs. In addition, the embedded test for board-level memories is usually treated as a sub-block embedded test structure. Chapter 3, "Logic Test and Diagnosis," describes the case of a scannable logic core that is treated as a sub-block.

By ensuring each sub-block's embedded test structure includes not only any required test controllers, but also scan access and isolation capabilities, it is much easier for the designer to make the rest of the chip's logic testable. The scan capability allows the surrounding logic to be tested without impacting the underlying non-scannable sub-blocks.

The Sub-Block Flow

The Sub-Block Flow comprises two steps that are run sequentially and that are repeated for each candidate sub-block in the design. The first step in the flow is to generate the embedded test structures that are required to test the sub-block. The result is a set of new HDL design objects that implement the embedded test. The next step is to create testbenches that verify the operation of these embedded test structures in isolation with the sub-block.

Figure 4-5 on page 100 illustrates the Sub-Block Flow.

Figure 4-5 The Sub-Block Flow

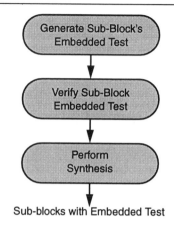

Sub-blocks with Embedded Test

Each step in the Sub-Block Flow is described in more detail in subsequent sections, beginning with the "Generating Embedded Test for Sub-Blocks" section.

Figure 4-6 illustrates a high-level view of the design hierarchy before and after the sub-block flow. The original sub-block contains the functional design. The tasks described in the next section add embedded test structures to this sub-block.

Figure 4-6 Architectural View of the Sub-Block Phase of the LogicVision Design Flow

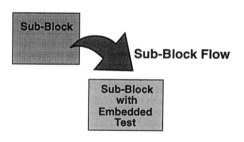

Generating Embedded Test for Sub-Blocks

The purpose of this step is to generate the embedded test structures for sub-blocks in the design that have been identified as candidates for their own embedded test capabilities. Typically, three types of design objects are generated for embedded test:

- Test collars that provide access to and isolation of the sub-block under test

- An embedded test controller for BIST of the sub-block

- An assembly that combines the sub-block, collars, and controller into a unit for ease of verification

In order to generate the embedded test structures for a sub-block, the following tasks must be completed:

1. Create a configuration file to serve as input to the Generate tool. This input file describes the properties and features to incorporate into the embedded test structures.

2. Run the Generate tool, specifying runtime options that govern how the tool operates.

The advantage of the Generate tool is that it can create an RTL description of the embedded test structures that the designer considers optimal for the sub-block. By selecting appropriate parameters in the configuration file, the designer has complete control over what embedded test features are generated to achieve an optimal test for the sub-block in the smallest possible area.

The runtime options allow the designer higher level choices, such as the type of HDL language to generate, and global characteristics of the design, such as clocking. The Generate tool creates the following output files:

● Log file

● RTL code of the embedded test structures

● Synthesis scripts for use with leading synthesis tools

● Simulation testbenches (Verilog or VHDL)

● Simulation scripts for use with leading simulation tools

Figure 4-7 illustrates the operation of the Generate tool.

Figure 4-7 Generate Tool Operation

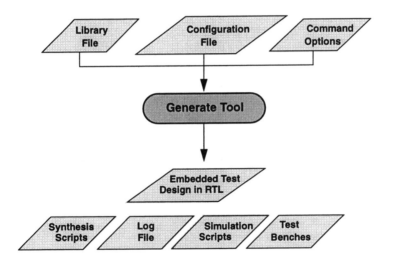

To control the results generated, the designer should always modify the input files and runtime options. The designer should *not* modify the output from the Generate tool directly because this impacts the repeatability of the results if changes are required. Modifying the output can also break the down-stream automation tools that rely on certain data to operate properly.

Verifying Embedded Test for Sub-Blocks

The purpose of this step is to verify the embedded test structures that have been generated in the context of the sub-block design object. This verification is performed to catch any errors or problems with the embedded test for the sub-block as early as possible. This verification is run on the assembly module that was created by the Generate tool using the testbenches output by the Generate tool and other testbenches that can be created using the Verify tool.

The following tasks must be performed to verify the operation of the embedded test for a sub-block: During verification, you check the sub-block's embedded test structures in two modes: basic modes of operation and programmable modes of operation, including specific test and diagnostic modes.

1. Simulate the testbench output by the Generate tool. During this simulation, the basic mode of operation of the sub-block's embedded test is verified.

2. Create a configuration file that serves as input to the Verify tool. This tool creates the additional testbenches to check specific test and diagnostic modes of the embedded test for the sub-block.

3. Run the Verify tool, specifying runtime options that govern how the tool operates.

4. Simulate the testbenches output by the Verify tool. During this simulation, specific test and diagnostic modes of the sub-block's embedded test are verified.

The above simulations are performed at RTL, before synthesis, and again at the gate level, after synthesis, to verify the timing correctness of the sub-block's embedded test. The Verify tool creates the following output files:

●   Log file

●   Simulation testbenches (Verilog or VHDL)

●   Simulation scripts

●   Information files used by other LogicVision tools

Figure 4-8 illustrates the operation of the Verify tool.

Figure 4-8 Verify Tool Operation

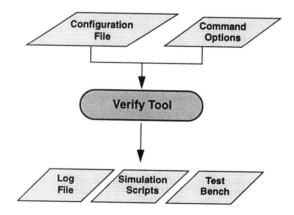

When the designer is satisfied with the performance of the embedded test for a given sub-block, and all sub-blocks have had embedded test added and verified, the designer can proceed to the next phase of the embedded test design flow—preparing the top-level logic block.

Preparing the Top-Level Logic Block

In this phase of the LogicVision design flow, the top-level logic block, which typically contains sub-blocks to which the embedded test was added in the previous phase, is made test-ready for scan and logic BIST. This phase is also referred to as the Prepare Logic flow. The preparation entails integrating the embedded test structures for the sub-blocks into the logic block. It also entails identifying and fixing design features that are incompatible with scan and logic BIST and adding embedded test structures to the logic for support of scan and logic BIST. During this phase, initial fault coverage calculations are also performed, and scan chains of the test-ready logic block are verified.

The Logic Flow

Figure 4-9 illustrates the design flow used to prepare the logic block. Each step in the Sub-Block Flow is described in more detail in subsequent sections, beginning with the "Integrating the Sub-Block's Embedded Test" section.

Figure 4-9 The Prepare Logic Flow

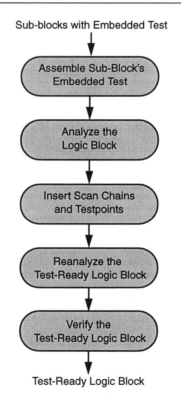

Figure 4-10 on page 104 is a high-level view of the design hierarchy before and after the Prepare Logic Flow. The top-level sub-block contains the functional design and any sub-blocks to which embedded structures were added as described in the "Adding Embedded Test to Sub-Blocks" section on page 95. The tasks described in the next section insert any embedded test structures for sub-blocks into the top-level logic block.

Figure 4-10 Architectural View of the Logic Prepare Phase of the LogicVision Design Flow

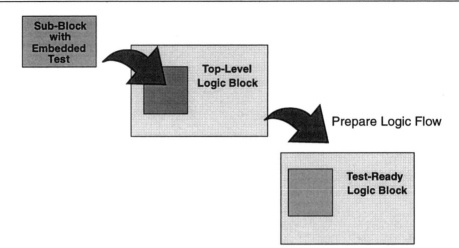

Integrating the Sub-Block's Embedded Test

The purpose of this step is to insert any embedded test structures for sub-blocks into the top-level logic block. This step involves replacing the sub-block instances with their *test-collared* versions; that is, versions that have test access circuitry added to their peripheries. The step also involves instantiating any embedded test controllers used to test the sub-blocks and routing the access and control ports of these controllers to the ports of the top-level logic block.

The following tasks must be completed in order to integrate the sub-block's embedded test into the logic block:

1. Create a configuration file to serve as input to the Assemble tool. This input file describes the instances to replace and the ports to connect.

2. Run the Assemble tool, specifying the runtime options that govern how the tool operates.

The Assemble tool uses information specified in a configuration file to determine where to replace existing modules with their collared counterparts and where to instantiate any embedded test controllers. The tool merges the test-collared sub-blocks and the embedded test controllers into the original design by writing a new file that contains a modified version of the design file. The original design files are left intact. The output of the Assemble tool consists of the following:

● Log file

● Top-level logic block design files modified to contain all of the sub-block's embedded test structures

● Release scripts that swap the modified design files and the original design files

Figure 4-11 on page 105 illustrates the operation of the Assemble tool.

Figure 4-11 Assemble Tool Operation

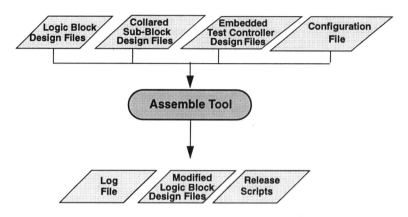

The designer can decide if this step is performed on the RTL view of the top-level logic block or on the gate-level view of the top-level logic block. If the step is performed on the RTL view, the logic block is synthesized later and then used in subsequent steps of this phase of the design flow. If the step is performed on the gate-level view, the modified logic block can be used directly for the rest of this phase.

A key benefit of the Assemble tool is that it allows all test-collared sub-blocks and related embedded test controllers to be inserted into the top-level logic block in a single step without manual intervention. This flexibility greatly reduces the risk of design time impacts and design errors.

Analyzing the Top-Level Logic Block

The purpose of this step is to investigate and improve the testability of the top-level logic block. Based on analysis of the logic block, certain design rule violations might require corrections. In addition, the number of test vectors applied through scan, or by the logicBIST controller, might require adjustments to achieve a desired fault coverage.

The Analyze tool used in this step runs on the gate-level description of the design because knowledge of the design's circuit-level implementation is needed to perform detailed analysis or to calculate fault coverage. The following tasks are typically performed during the Analyze step:

1. Extract the clock domains to understand how to make scan test and logic BIST robust between these domains.

2. Perform design rule-checking to ensure that the design does not contain any elements that are incompatible with the scan or logic BIST test methodologies and tools. For example, the design rules ensure that all logic is synchronous and that tri-state bus drivers have fully decoded control signals.

3. Identify all redundant faults. A *redundant fault* is a fault that cannot be detected with a test vector. A large number of redundant faults reduces the fault coverage and usually indicates an incorrect use of signals held at a constant value to fix scan rules violations.

4. Decide if testpoints are required. *Testpoints* consist of special cells that are inserted into the netlist to improve controllability or observability of difficult-to-test nets. The Analyze tool locates nodes that are resistive to test with BIST patterns and models the improvement in fault coverage that results from the use of testpoints at these nodes.

The designer must fix any violations of these rules by modifying the logic block. The information extracted from the design using the Analyze tool is used by other tools in the design flow, including the Scan Generate tool, which is used for scan and testpoint insertion in the next step of the flow. Figure 4-12 illustrates the operation of the Analyze tool.

Figure 4-12 Analyze Tool Operation

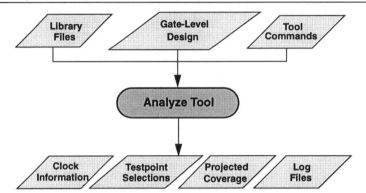

The designer chooses the number of testpoints to add based on the trade-off between fault coverage and area overhead. The number of testpoints needed to achieve a given fault coverage varies from one design to another. However, as a general rule, achieving a fault coverage that is close to 100 percent using logic BIST typically requires approximately one testpoint (about 10 gates) per 1000 gates of logic.

The Analyze tool enables analysis of the trade-off between number of testpoints and fault coverage. The tool can be constrained to *not* place testpoints on specified critical paths in the design. This option can be important because testpoints implementing controllability add a gate delay to the path on which they are inserted.

One way to reduce the number of testpoints is by limiting the fault coverage that must be achieved with logic BIST. The remaining fault coverage can be obtained through scan testing using a relatively small set of deterministic scan vectors generated by ATPG.

Inserting Scan and Testpoints

During this step, scan chains, and possibly testpoints, are added to the top-level logic block. Adding scan chains to the logic block involves substituting scan flip-flops for non-scan flip-flops and connecting these scan flip-flops into one or more scan chains. Grouping scan flip-flops into scan chains and ordering the flip-flops within each chain depends on several factors:

- *External scan test*—For external scan test, the total number of independent scan chains is limited by the number of chip pins that can be used as scan inputs and scan outputs. This number is typically dictated by the scan channels available on the tester used to apply the scan tests. Maximizing the number of scan pins and, therefore scan chains, maximizes the scan throughput and thus minimizes test time.

- *Logic BIST*—The logicBIST controller drives the scan chains rather than an external tester. Because the constraint on scan pins is eliminated, it is possible to create a large number of reduced-length scan chains that are driven directly by the BIST controller. This feature greatly reduces test times because the time required for scan shifting is greatly decreased.

- *Layout*—The order of the scan flip-flops within a scan chain influences the operation of the physical layout tools. Often, the scan-chain order is revised by the layout tools to achieve optimum layout of the functional design.

- *Clock domains*—Scan flip-flops in different clock domains must be placed in separate scan chains. It is also necessary to equip scan flip-flops on clock domain boundaries with special capture-by-domain capabilities in order to handle timing issues across these boundaries during scan and logic BIST testing.

The Scan Generate tool inserts scan chains hierarchically on the entire design or incrementally block by block. When adding scan chains to a given block in the design hierarchy, any existing scan chains in the sub-blocks are merged into the scan chains in the higher-level block. Chains are automatically balanced by clock frequency to optimize test time.

The Scan Generate tool can add the testpoints, which were selected in the previous Analyze step, to the design either at the same time that scan chains are added or in a separate operation.

The input configuration file for the Scan Generate tool provides user information about the desired names of the ports on the scan chain and any existing scan chains in the design. The tool's runtime options are used to control trade-offs such as chain length and number of chains as well as the search path for the design files. The output from the Scan Generate tool consists of the following:

- Log file
- The scan and/or testpoint-inserted design
- Release scripts for swapping the new and old design versions

Figure 4-13 illustrates the operation of the Scan Generate tool.

Figure 4-13 Scan Generate Tool Operation

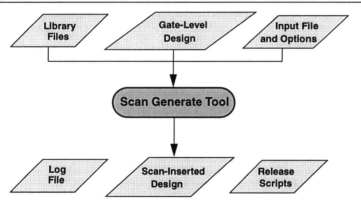

LogicVision's Scan Generate tool is unique in its ability to automatically implement a timing-robust, capture-by-domain scan methodology. This methodology ensures that cross-domain scan timing issues are resolved and that the scan chains can be used for at-speed logic BIST.

Reanalyzing the Test-Ready Logic Block

After adding the embedded test for the sub-blocks and the scan chains to the top-level logic block, it is necessary to reanalyze the top-level logic block. This reanalysis ensures that the sub-block's embedded test is correctly integrated, scan chains are properly connected, and design changes do not cause any additional violations.

The Analyze tool is rerun on the test-inserted, top-level logic block. During reanalysis, the Analyze tool performs the following tasks:

1. Checks that the design is free of rule violations and determines that the scan chains and testpoints are correctly inserted.

2. Performs a fault simulation to determine the actual fault coverage of the logic block.

3. Verifies that the sub-block's embedded test are properly connected and extracts the test connections map for use in the next step.

During reanalysis, the Analyze tool generates several output files:

- Log files, which include the coverage information

- Files that list all untested gates, unused gates, and unused ports

- A few random scan vectors used in the next step for verifying scan

- The test connections map file, which describes all embedded test features

Any errors flagged in the log file must be corrected and the tool must be run again. When the design is error-free, the Analyze tool can complete the rest of the operation. Figure 4-12 on page 106 illustrates the operation of the Analyze tool.

Now the logic block is test-ready. The final step is to verify that the scan chains, testpoints, and other embedded test structures work correctly.

Verifying the Test-Ready Logic Block

The final step in the Prepare Logic Flow entails simulating the test-ready logic block that contains the sub-blocks embedded test, scan chains, and any optional testpoints. This simulation verifies that the sub-block's embedded test still functions properly, scan chains and testpoints operate at speed, and the functionality of the original logic block is unaltered.

The designer should rerun the verification of each sub-block's embedded test in the context of its integration into the complete, test-ready logic block. This verification is performed exactly as it was in the Sub-Block Flow, except that the Verify tool is rerun using the test connection map file created by the Analyze tool when it is rerun on the top-level logic block. Refer to the section "Verifying Embedded Test for Sub-Blocks" on page 101 for a detailed description of the verification.

To verify that the scan chains and testpoints are operating correctly, the Verify tool is used to create a testbench for use with simulation. The files that must be provided as input to this verification tool include the following:

- A file that contains the scan patterns to simulate

- A file generated by the Analyze tool that provides the scan chain ordering and scan I/O pin information

Based on these input files, the Verify tool creates a testbench used to simulate the application of the scan vectors to the test-ready logic block. Typically, it is sufficient to simulate only a small number of the scan patterns to validate that the scan chains and testpoints have been correctly inserted in the top-level logic block. The scan verification tool generates the following output files for reference and simulation.

- Log file

- Verilog or VHDL testbench that drives the simulation of the pseudo-random scan patterns to the top-level logic block

Figure 4-14 on page 109 illustrates the operation of the Verify tool.

Figure 4-14 Verify Tool Operation

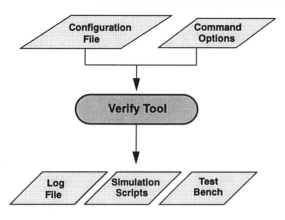

The simulation time in a large design of even a few scan cycles can be significant, particularly if simulation is performed using a testbench that shifts the scan chain each clock cycle. The simulation time can be reduced by eliminating some of the serial scan cycle by using the concept of parallel-load scan chain simulation.

Parallel-load simulation thoroughly exercises the logic block while saving the time required to shift each vector into the scan chains. The parallel-load simulation scheme follows this approach:

1. Use a *force* command to load the scan chains with the data that they would contain after the *next-to-last* shift clock.

2. Remove the *force* condition.

3. Apply the last shift clock.

4. Apply the capture clock.

5. Compare the contents of the scan chains with the expected contents.

Figure 4-15 and Figure 4-16 on page 110 illustrate the parallel-load simulation scheme. The *force* command (force "1" or "0") is applied directly to the scan data input of each flip-flop of each scan chain. On the next positive edge of the clock, the force data is clocked into the flops. Next, the force condition is released. On the next positive clock edge, the last shift occurs. The last positive edge is the data capture clock. The simulator compares the captured data in each flop to the expected result.

Figure 4-15 Waveforms for Parallel-Load Simulation

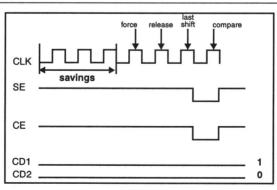

Figure 4-16 Parallel-Load Forcing Scheme

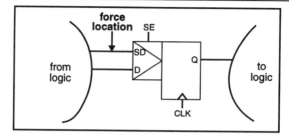

Any problems that are encountered during simulation must be identified and corrected before proceeding to the next step in the design flow. When the designer is satisfied with the performance of the embedded test and scan chains within the test-ready logic block, the designer can proceed to the next phase of the embedded test design flow.

Adding Embedded Test to the Top Level of the Chip

In the first two phases of the LogicVision design flow, the designer adds embedded test to sub-blocks and makes the top-most logic block ready for test using scan. (Refer to the "Adding Embedded Test to a Sub-Block" section on page 99 for the first phase and the "Preparing the Top-Level Logic Block" section on page 103 for the second phase.) The final phase entails adding embedded test infrastructure to the top-level of the chip. In this phase, the designer uses the LogicVision tools to generate top-level embedded test structures, assemble these structures into the chip, and connect the top-level embedded test to the embedded test structures in the test-ready logic block. Finally, the entire chip design with embedded test is verified and all the test patterns for manufacturing are output.

Top-Level Flow

The designer includes the remaining elements of the embedded test architecture at the top level of the chip by completing the tasks illustrated in Figure 4-17.

Figure 4-17 Top-Level Flow

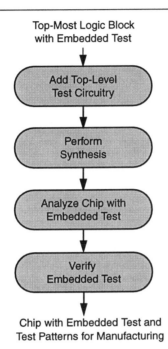

Top-Most Logic Block
with Embedded Test

Add Top-Level
Test Circuitry

Perform
Synthesis

Analyze Chip with
Embedded Test

Verify
Embedded Test

Chip with Embedded Test and
Test Patterns for Manufacturing

The Top-Level Embedded Test module in Figure 4-18 on page 112 generally contains the following elements:

- IEEE 1149.1 TAP controller—operates as the embedded test manager
- Boundary scan register—enables control of pad cells for DC parametric chip testing and board-level interconnect test
- Logic BIST controller—provides complete at-speed self-test of the logic
- Clock prescaler—provides clock speed and power management
- Scan-chain router module—configures the scan chains for logic BIST or scan test
- At-speed interconnect test controller—used for testing board-level interconnects via boundary scan.
- Fault insertion cells—for use with boundary scan in system test.
- Connections to all the embedded test features in the test-ready logic block

The Chip Design block contains the designer's pad cells.

Figure 4-18 illustrates a high-level view of the design hierarchy before and after the top-level flow.

Figure 4-18 Architectural View of the Top-Level Phase of the LogicVision Design Flow

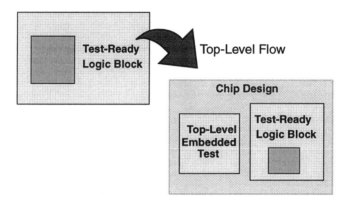

Adding Top-Level Embedded Test

This section describes how the designer generates the top-level embedded test structures and merges them into a design. During this step of the LogicVision design flow, the designer adds a TAP controller, boundary-scan register, scan-chain router, and logicBIST controller to the top level of the chip and connects these design objects to pins of the chip and ports for embedded test on the test-ready logic block.

The TAP controller, which is compatible with the IEEE 1149.1 standard, can be configured to support any number of embedded test features within the chip. As previously mentioned in the "Logic BIST Building Blocks" section on page 69, the TAP controller acts as test manager, initializing and running all embedded test controllers based on instruction register operation codes.

Input files that reflect the test specifications are prepared by the designer and read by the Top-Level Integration capability. The Top-Level Integration capability combines a Generate tool and an Assemble tool that are used consecutively to generate and merge the top-level embedded test.

Information that the designer specifies in the input file includes the following:

● Logic test configurations to implement

● Assignment of boundary-scan cell types to top-level signals

● Order of cells in the boundary-scan register

● Requirements for the logicBIST controller and TAP controller

● Connections between ports of the TAP controller and test ports for embedded test of sub-blocks within the test-ready logic block

Based on this input file, and various runtime options, the Top-Level Integration capability creates synthesizable RTL descriptions of the top-level embedded test structures. This capability then merges the embedded test structures into the top-level design description, making all necessary connections to complete the embedded test implementation for the entire chip.

Figure 4-19 on page 113 illustrates the top-level integration.

Figure 4-19 Top-Level Integration

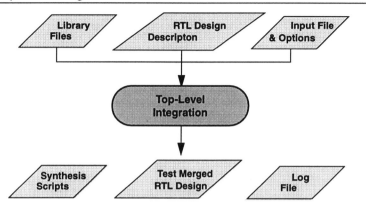

The designer can perform this step on either the RTL or gate-level design. When top-level integration is performed on the RTL design, the designer achieves a complete RTL model of the entire chip, including embedded test.

After performing top-level integration, the designer must synthesize the top-level embedded test structures and any remaining RTL blocks for the top level of the chip before beginning the next step in the process. The synthesis scripts for the top-level embedded test structures can help simplify this effort.

Analyzing the Top-Level Embedded Test

After using the top-level integration tool to add the top-level embedded test to the chip, the designer must analyze the entire test-merged chip. The following tasks must be performed:

1. Ensure that the top-level embedded test structures comply with all design rules for scan test and logic BIST.

2. Obtain constraint files that describe the implemented test configurations for static timing analysis.

3. Create automatic test pattern generation (ATPG) scan vectors used with the multi-chain scan test configuration, if necessary.

4. Obtain the reference signature for the logic BIST *default* test mode. This signature is used to create the hard-wired signature module that is included within the logicBIST controller. The signature module does not require synthesis because it consists only of connections to design nets with static 1 or 0 logic values.

5. Obtain the final fault coverage for logic test. The designer should already have a very close approximation of the fault coverage based on the work completed while making the top-level logic block test-ready.

6. Extract a description of the embedded test architecture for the entire chip from the chip's gate-level netlist, check extracted information against connection rules, and generate test connection map files for the top-level Verify step.

These tasks are performed automatically using the Analyze tool on the gate-level description of the of the chip. Figure 4-20 on page 114 illustrates the Analyze tool operation.

Figure 4-20 Analyze Tool Operation

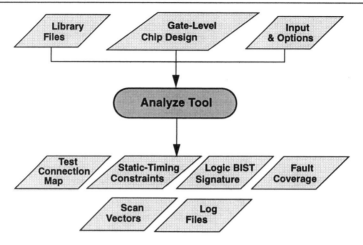

The designer must carefully review the output of the top-level Analyze tool run for any errors and correct the errors before proceeding. Errors might have been introduced because of incorrect input to the Top-Level Integration tool during the preceding step in the flow. If these errors are caught before top-level verification, verification time can be reduced.

Verifying Embedded Test from Chip Pins

In the final step of the LogicVision embedded test design flow, the designer performs tasks to verify the functionality and timing of the all the embedded test capabilities in the entire chip. When this work is complete, test patterns for chip manufacturing test are produced. The following is a summary of the tasks performed during this step.

1. Create simulation testbenches to verify the functionality of each portion of the embedded test architecture.

2. Run gate-level simulation for each embedded test structure using the testbenches.

3. Perform static timing analysis on the entire chip to verify timing of the embedded test.

4. Create test patterns for manufacturing in WGL or SVF format.

The Verify tool provides the designer with considerable flexibility. Using the tool, the designer can make trade-offs between embedded test runtime, power, resolution, and coverage. The designer prepares a Verify tool input file for each of the embedded test structures that have been integrated into the chip, specifying the following test setup information:

● Frequency of operation of the clock driving the embedded test

● Configuration and number of test patterns to use

● Test operations to perform and their sequencing

● Test mode (default, normal or diagnostic) of the embedded test

● Parameters to measure (for example, in mixed-signal BIST)

The verification testbenches output by the Verify tool at the top level provide operation codes for the instruction register of the TAP controller to establish the appropriate, user-defined test setup for each different embedded test capability. Using these testbenches, the designer can simulate each of the embedded test capabilities at the gate level.

Static timing analysis should be performed on the entire chip, using the timing constraint information output by the Analyze tool, to confirm the timing robustness of the embedded test of the chip.

For logic BIST and scan test, the designer normally simulates the functionality of a few vectors at the beginning and end of the test using the parallel-load capability described on page 108. This simulation, together with static-timing analysis of the entire chip, is sufficient to validate the correct performance of the logic BIST and scan test.

Figure 4-21 illustrates the Verify tool operation at the top level of the chip.

Figure 4-21 Verify Tool Operation

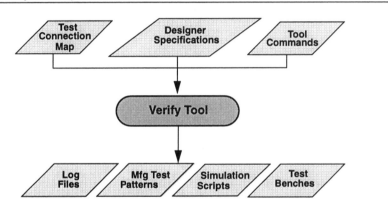

The test patterns can be output simultaneously with the creation of the testbenches or output later after each of the embedded test features has been verified. Test patterns can be output in either WGL format for use on chip testers or in SVF format for use on board testers. It is generally useful to output both a *go/no-go* test and a more detailed diagnostic test for each embedded test capability for shipment to manufacturing.

Chapter 5
Hierarchical Core Test

VDSM (0.25 microns and below) technology provides the potential for designers to create SOC devices. However, to realize this potential and meet time-to-market requirements, designers must use new approaches to chip design. It is commonly recognized that, in order to complete designs comprising millions of equivalent gates within a satisfactory timeframe, designers must reuse pre-existing, application-specific functional blocks or *cores*. These types of blocks are available from many sources including internal company libraries, semiconductor manufacturers, and third-party IP providers.

Soft cores are described in RTL format, and hard, or legacy, cores are typically *design blocks* that are available only as physical layouts targeted to a specific semiconductor manufacturer's technology.

Unfortunately, many cores currently do not contain any provision for test other than a set of functional patterns. Functional patterns often do not provide satisfactory fault coverage for manufacturing test and are seldom suitable for diagnostic work by designers or test engineers. If a soft core is designed to adhere to a few basic synchronous design rules, the core can be easily fitted with a suitable structural test, either scan test or BIST.

Testing Cores

Use of embedded cores divides the IC design community into two groups: *core providers* and *system chip integrators*. In traditional System-on-Board (SOB) design, the components that go from the provider to the user are ICs that are independently designed, manufactured, and tested. Users of these components are concerned only with the design, manufacturing, and testing of their systems. With respect to testing, they can assume fault-free components, and hence the testing is reduced to manufacturing defects in the interconnect between components. In SOC design, the components are cores. Because the cores are not yet manufactured or tested, the integrator is responsible for manufacturing and testing the entire chip, including the logic within the cores themselves. This chapter describes a two-step design process for designing system chips using embedded cores.

Hierarchical SOC Design Process

Given that system chips are designed using reusable cores, the test methodology should accommodate and facilitate this design process. In fact, to meet the objectives of *plug-and-play* core reuse to significantly enhance designer productivity, it can be argued that the core test must also be reused. This implies that after a core has been designed, its associated test can be generated and verified without knowing in which application it will be embedded. The design process in Figure 5-1 on page 118 inherently supports core test reuse.

Figure 5-1 SOC Design and Test Process

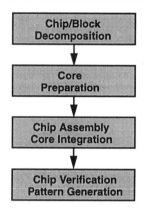

The test process follows the design process illustrated in Figure 5-1.

1. Chip or Block Decomposition

    The first step in the design process is to decompose the chip or block into identifiable blocks or modules. This decomposition process involves up-front planning to identify all significant functional blocks. The functional blocks can then be acquired either from third parties or as part of a corporate internal library, or the blocks can be commissioned as a new core development activity. Decomposition can also be undertaken to divide a very large design into more manageable pieces for design verification and test purposes. The decomposition process identifies all prerequisites for moving forward with the design and thus is an important aspect of the system-chip design process.

2. Core Preparation

    This step ensures that the core is fully functioning and ready for chip integration. The core is classified as either *test mergeable* or *test non-mergeable*. Test-mergeable cores are tested as part of the system chip and must be made *test ready*. Test non-mergeable cores are tested as a separate entity at the chip level and thus require *access* and *isolation*. The full benefit of hierarchical test is achieved when all cores are designed as test non-mergeable. This ensures that internal core test is addressed up front during core design and not left as an afterthought to the chip integrator.

3. Chip Assembly and Core Integration

    Here, the cores and any UDL are brought together to define the complete design. Chip assembly also involves optionally adding IEEE 1149.1 TAP and boundary-scan logic. Chip-level test requirements usually dictate whether a top-level logicBIST controller is used to test the UDL and interconnect between cores. Alternatively, this logic can be tested using scan ATPG techniques.

4. Chip Verification and Pattern Generation

    Both functionality and manufacturing test are verified. Because significant blocks of the design can include their own embedded test resources, chip-level test verification largely involves testing the interconnect between cores and accessing their embedded test resources for core internal test.

The process can be repeated by considering the assembled chip as a core to reuse in a more complex application.

Embedded Core Test Challenges

Testing complex SOC designs involves all the inherent difficulties associated with very large ASICs as well as the difficulties of incorporating reusable cores. It is often necessary to separate the issues of dealing with large complex ASICs from the core reuse issues. The large ASIC test issues must be dealt with in any case, and can generally be considered a subset of the more complex problem. Thus the SOC test solution for reusable cores can be applied equally to large ASICs. This chapter focuses on the test challenges of reusing embedded cores. Several factors require consideration: the internal core logic must be tested; the core might be designed by a third party that is unable or unwilling to divulge the complete structure; and the core might be used in multiple applications with different test requirements. The solution must address component production test, diagnostics, and failure analysis—and enable test reuse in the board and system. At-speed testing is often considered essential for ensuring high quality. Given the requirement for test reuse, it is evident that BIST plays a prominent role in the architecture. The issues associated with testing core-based system chips can be categorized in four ways:

- **Core-internal tests.** Given the diversity of core types and test quality goals, it is anticipated that core internal test might vary between core providers. One core provider might provide functional vectors to be applied through chip pins, another might provide reusable ATPG scan vectors, and a third might provide BIST within the core. The architecture should support each of these scenarios.

- **Core test pattern transfer.** Because the core provider might be from a separate group or company, the chip integrator must be able to verify and generate the chip manufacturing test programs from information delivered along with the core. Information about such items as core internal DFT, test modes and corresponding test protocols, fault coverage, and test patterns must be provided in easily reusable formats.

- **Access to embedded cores.** Because the core test patterns provided by the core developer usually refer to the core terminals, and the core can be deeply embedded within the system chip, a test access mechanism (TAM) must be provided to link the test pattern *source* to the core inputs and the core outputs to the test pattern *sink*. The source and sink can be either on chip (BIST) or off chip (external ATE). The interface to the core terminals is usually provided by a *test collar*. This test collar also provides isolation to ensure that each core internal test does not depend on surrounding logic. The collar also provides a way for testing the interconnect between cores or UDL that surrounds the core (external test).

- **System-chip test integration.** The overall system-chip test should cover the embedded cores, the UDL, and the interconnect logic between cores. Flexibility should be allowed for the chip integrator to make trade-offs regarding test quality, test time, power consumption, and area.

Hierarchical Embedded Core Test Architecture

A generic test architecture for supporting embedded cores is shown in Figure 5-2 on page 120. The architecture consists of the following:

- A source—generates test stimuli
- A sink—evaluates the test responses
- A TAM—transports the stimuli from source to core and from core to sink
- A collar—connects the TAM to the core

The source and sink can optionally be off chip (external ATE testing), on chip (shared central test resource), or within the collar (dedicated test resource). For an autonomously self-testable core that can be reused in multiple applications, it is often preferable to include the source and sink within the collar. Including the source and sink in the collar ensures that the same level of test quality is achieved for each application and minimizes the amount of test preparation and verification at the chip level. Only then can true plug-and-play core reuse be achieved.

Note: To be consistent with other chapters in this book, the term **collar** *refers to the logic surrounding the core that connects the TAM to the core terminals. The term* **wrapper** *is sometimes used to describe the same concept.*

Figure 5-2 Hierarchical Embedded Core Test Architecture

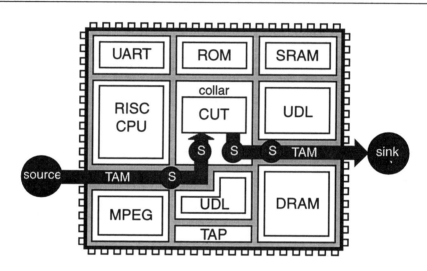

Collar Functions

The collar is a thin layer of logic that surrounds the embedded core and provides a switching mechanism to support the various access mechanisms. As part of the functionality of the collar, a collar element that provides the following capabilities is associated with each core terminal.

For each core *input* terminal:

● A normal mode where the core terminal is driven by the host chip

● An external test mode where the collar element observes activity on the core input terminal used for interconnect and UDL testing

● An internal test mode where the collar element controls activity on the core input terminal used for internal core testing

For each core *output* terminal:

● A normal mode where the host chip is driven by the core terminal

● An external test mode where the host chip is controlled by the collar element used for interconnect and UDL testing

● An internal test mode where the collar element observes core terminal activity

As mentioned earlier, in the section "Hierarchical Embedded Core Test Architecture" on page 119, the source and sink of the patterns is optional and can be within the collar, within the chip, or external to the chip. Typical configurations include the source and sink within the collar for internal testing or the source and sink within the chip for external (interconnect or UDL) testing. Two typical exceptions to this rule are embedded core characterization support, where access to the core terminals from chip pins is provided (external ATE source and sink), and core diagnosis, where core internal scan chains are accessed from chip pins (external ATE sink only).

Collar Generation

Collar generation can be performed by the core provider or the system chip integrator. To obtain the full benefits of improved designer productivity, it is recommended that the core provider generate the collar. This task partitioning is especially compelling when the core will be used in many applications. However, there are benefits in having the core designer generate the collar—even in situations where the core will be used in a single application. The core provider implements some collar functions by sharing functional logic; sharing logic is rarely possible when the system chip integrator provides the collar. The core provider also has knowledge of the core internals and is better able to assess testability issues and thus develop a scheme that facilitates core internal testing. It is anticipated that, in a true core-based design, the core external test will largely be point-to-point interconnect testing, which usually does not require special consideration. The core provider should ensure that all likely applications for this core are addressed during collar generation. In particular, special consideration should be taken to deal with any data signals that cross clock domains. Robust techniques are available that allow designers to handle timing considerations early in the design process.

Hierarchical Embedded Core Test Solution

This section describes a particular solution to the embedded core system chip test problem. It implements the concepts described in the previous sections and includes many architectural features currently under discussion as potential industry standards. In particular, two groups deal with standardization issues for core test access: IEEE P1500 Standards for Embedded Core Test and Virtual Socket Interface Alliance (VSIA). It is anticipated that there will be a migration to full compliance after the standards have been ratified. There will likely be a phase-in period during which both the present solution and compliant cores will be integrated into a single device.

The solution supports the automated creation of test infrastructure for several types of cores, including the following:

- Cores with functional patterns that cannot be modified (commonly referred to as *legacy cores*)
- Scannable cores that will be merged and tested with other chip-level logic
- Scannable cores that will remain separate test entities where ATPG patterns will be reused at the chip level
- Cores planned for BIST implementation that will be merged and tested with other chip-level logic
- Cores, in which BIST has been implemented, that will remain separate test entities where an embedded controller performs the test

These capabilities are commonly used by both core providers and chip designers who use cores to implement their SOC designs. Core providers either integrate test infrastructure into their cores or certify that their cores are compatible with a certain test mechanism that will be implemented by the chip designer. The embedded test logic allows the chip designer to add the test-prepared cores as plug-and-play modules to a design. The chip designer can also use certified test-ready cores, knowing that test-related issues have been dealt with by the core provider. The chip designer also uses the same capabilities to automatically create chip-level test features and access to embedded test resources. The solution is fully scalable in that the chip designer can easily package the chip as a core to be reused in a more complex application.

Figure 5-3 illustrates the chip-level view of the hierarchical embedded core test solution. In this example, the chip integrator has decided to include an IEEE 1149.1 TAP and boundary-scan register. The figure shows a source and sink within the circuit-under-test (CUT) collar, which represents an embedded logic BIST controller. The thin arrow that connects the CUT collar to the TAP represents the control signals that provide the interface for initiating the test and monitoring status. The thick black lines represent the TAM used for external testing. Here, either ATPG patterns can be applied to the CUT or, optionally, functional patterns can be applied from chip pins. The TAM for external access is often not necessary for cores that include dedicated test resources. Often this TAM interface is provided on the core, assuming that the chip integrator will decide whether or not to use the interface. The cost of including the TAM interface on the core is usually quite negligible.

Figure 5-3 Chip View of Embedded Core Test Architecture

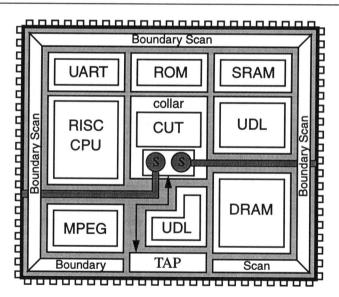

Within the SOC test architecture, the various cores, memories, and mixed-signal blocks are first tested in the following ways:

● Legacy cores are isolated and accessed from package pins for functional test with patterns from the core provider. This capability can also be used as an option for cores that also include DFT support such as scan or logic BIST. For example, by providing chip-level access to the core terminals, a chip integrator can characterize a core in a new process technology.

- Fully scannable cores are isolated from surrounding logic and accessed from package pins for scan test. Alternatively, the scannable logic can be fully tested by a logicBIST controller that is either embedded within the *test-collared core assembly* by the provider or within the chip by the chip integrator.

- Embedded memories, either within a core or in UDL, are accessed and tested by embedded memory BIST controllers.

- Mixed-signal blocks, such as PLLs, are accessed and tested by embedded BIST controllers.

After the individual cores are tested, chip-level testing is performed. UDL not included in a core and the connections between cores is tested with a *global* scan test or logic BIST. Thus, the system chip is tested in two phases. First, core internal tests are performed, followed by core external tests to address the interconnect and UDL. The hierarchical test sequence is managed through a designer-customized TAP controller. Simple instructions from external hardware to the TAP controller select one or more test operations to perform. Several memories or cores can be tested simultaneously, yielding a significant reduction in test time. Full flexibility is provided for scheduling concurrent testing; the trade-offs are test time versus power consumption. After the selected test operation is completed, another test can be initiated with a new instruction to the TAP controller.

The embedded test logic allows a designer or test engineer to debug the scannable logic within each individual core without outside influence from other cores or design blocks contained in an SOC design. In addition, UDL and connections between cores, memories, and the UDL can be debugged with the contents of the cores and memories ignored. The ability to treat the core contents and memories as *black boxes* at the chip level not only facilitates diagnosis but also makes the solution very scalable. It is rare that the whole system chip needs to be flattened for analysis. In fact, the core internals are often not available to the system chip integrator, therefore requiring support for black-box analysis.

Legacy Core Test Hardware

The support for legacy cores involves preparing a test collar for the core, merging the test-collared core into the host design, and generating an IEEE 1149.1-compliant TAP controller to schedule functional testing of the core from the chip pins. A Verilog or VHDL verification testbench to ensure that the chip-level test infrastructure is working correctly is automatically generated.

Preparation of Scannable Cores

Scannable cores can be prepared for scan test and logic BIST by inserting scan chains and, optionally, testpoints. These scannable cores will be test-merged later with other system-chip logic. The output from this flow is a core with inserted scan chains and testpoints that complies with all rules for a scannable design. Data files are also generated that allow scan-chain insertion at the chip level to automatically interface with the embedded scan chains.

To prepare scannable cores that will not be merged with surrounding logic, a test collar is generated. This test collar serves two purposes. The test collar isolates the scannable core from the surrounding logic (during internal test) and provides scan access for testing surrounding logic (during external test). The test collar is necessary to ensure that the generated ATPG scan patterns for testing the core internal logic are reusable for system-chip manufacturing.

To prepare scannable cores that will be tested with an *embedded logicBIST controller*, a test collar is generated. This test collar provides collar elements to access and isolate each embedded core terminal, as well as a logic BIST controller. The isolation is necessary to ensure that logic BIST is not influenced by surrounding logic since this core might be used in many applications. The collar also enables testing of surrounding logic.

In each of the above scenarios all generated hardware objects are provided in RTL form in Verilog or VHDL formats. Data files are generated for automatically merging test-collared cores into target designs and for performing scan-chain insertion at the next level of hierarchy. Simulation testbenches and manufacturing test patterns are automatically generated at both the test-collared core level and the chip level. ATPG patterns generated at the core level are directly reusable at the chip level.

Embedded Logic Test Architecture

This section describes in more detail the embedded logic test architecture. Capabilities are provided for supporting fully isolated reusable cores with built-in test resources. In fact, the legacy core and the scannable core support are subsets of the architecture, in that the logicBIST controller can be optionally excluded. The embedded logic test architecture includes the following:

- Three internal test configurations—single scan chain, multiple scan chains, and logic BIST, as described in the section "Internal Test Configurations for the Controller" on page 124

- An external test configuration—interface test for connections between cores, memories, and UDL (described on page 129)

- Core/configuration select mechanism (described on page 131)

- Collar elements (described on page 131)

Internal Test Configurations for the Controller

One of the major requirements of the embedded logic test architecture is support for internal core testing. Reusable cores can be of significant logic density and complexity to warrant particular attention and potentially dedicated test resources. The embedded logicTest controller can be configured in the following modes to address internal core testing:

- For ATPG scan test. In this case, scan patterns are applied from an external tester to the package pins of the design. By setting an appropriate instruction of the TAP, the embedded test infrastructure routes the scan patterns, scan-enable signal, and scan clock through the design hierarchy to the scan chains of the embedded core. Collar-element scan chains are also accessible from chip pins to provide access to the embedded core terminals and isolation from surrounding logic. The logicTest controller supports two scan test modes:

  - A single scan chain with the scan patterns applied from the TDI port of the TAP and the responses shifted from the TDO port. The scan chain includes the scan chains of the core, the scan segments of the core's test collar, and the scan chains of the embedded logicBIST controller (if included in the logicTest controller). This test arrangement is referred to as the *Single-Chain Internal Test* configuration. It is a mandatory configuration used for diagnostic purposes and silicon debug.

- Multiple scan chains with the scan patterns applied from the TDI port of the TAP and from package signal pins that are shared between functional mode and test mode. The responses are shifted from the TDO port of the TAP and from package signal pins. The scan chains include the scan chains of the core, the scan segments of the core's test collar, and the scan chains of the embedded logicBIST controller (if included in the logicTest controller). This test arrangement is referred to as the *Multi-Chain Internal Test* configuration. It is an optional configuration that is used for component manufacturing test as an augmentation to logic BIST.

- For logic BIST. An embedded logicBIST controller can apply pseudo-random patterns to the scan chains of the core and the core's test collar. Special consideration is taken to ensure hazard-free, at-speed test of a core with multiple clock-domains operating at multiple frequencies. This test configuration is referred to as the *Logic BIST* configuration and is an optional configuration.

Single-Chain Internal Test

In the Single-Chain Internal Test configuration, all scan chains of the test-collared core assembly are concatenated into a single chain, as illustrated in Figure 5-4. The single chain includes the following elements:

- Scan chains internal to the embedded core
- Scan chains of the test collar
- Scan chains of the embedded logicBIST controller (optional)

The single concatenated scan-chain input and output are generally connected directly to the chip-level TAP. In this case, the TAP must accommodate multiple user TDRs. By selecting the appropriate TDR of the TAP, each specific core can be individually accessed. This mechanism is used for initializing embedded controllers and monitoring status as well as for diagnostic purposes. For example, in diagnostic mode, ATPG scan patterns can be applied for slow-speed structural verification or logic BIST patterns can be diagnosed by shifting out the embedded core contents after application of BIST.

The connections to and from the chip-level TAP to the embedded core single scan chain are automatically performed. These connections are also structurally extracted and verified by a rules checker.

Figure 5-4 Single-Chain Internal Test Configuration

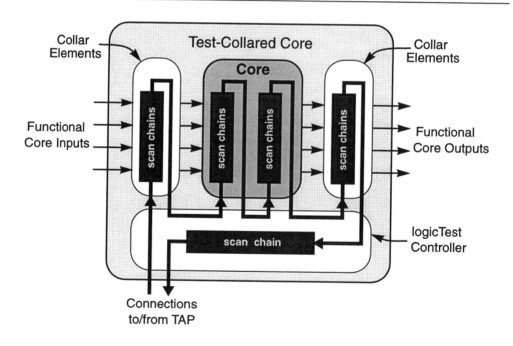

Multi-Chain Internal Test

In the Multi-Chain Internal Test configuration, all scan chains of the test-collared core assembly are concatenated into a user-specified number of scan chains. The resulting chains include the following elements:

- Scan chains internal to the embedded core
- Scan chains of the test collar
- Scan chains of the embedded logicBIST controller (optional)

Figure 5-5 shows the The Multi-Chain Internal Test configuration. The multiple chains are segmented into one scan chain that is directly connected to or from the chip-level TAP and an auxiliary set of scan inputs and outputs. The TAP scan-chain connection is identical to the connections used in the mandatory single-chain internal test configuration. The auxiliary scan connections are automatically routed through the hierarchy to chip-level pins. There they can be shared with functional pins or left as dedicated scan ports if additional device pins are available. The Multi-Chain Internal Test configuration is an optional configuration and is used to perform scan testing of the test-collared core. The ATPG patterns generated at the test-collared core level are automatically retargeted to chip pins; thus it is not necessary to regenerate ATPG patterns for the embedded core. This ensures that the pattern validation does not have to be performed multiple times for each application of the reusable core and can significantly enhance designer productivity as a result.

Figure 5-5 Multi-Chain Internal Test Configuration

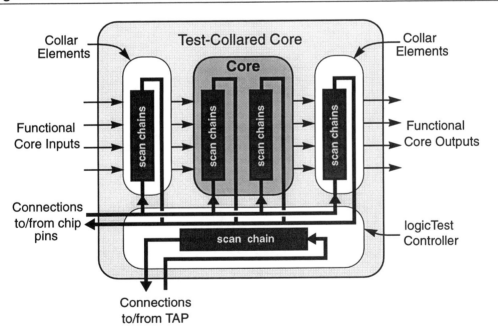

Logic BIST

In the logic BIST configuration, the collar-element scan chains and the core's internal scan chains are connected to the PRPG and the MISR within the logicBIST controller. Figure 5-6 shows the logic BIST configuration. The scan chain, used to initialize the logicBIST controller, is connected to a TDR of the TAP controller. Typically, the logic BIST configuration is used to perform at-speed testing of the test-collared core. This test configuration can easily be reused at board- and system-level with a few simple instructions to the TAP controller of the host chip.

Figure 5-6 Logic BIST Configuration

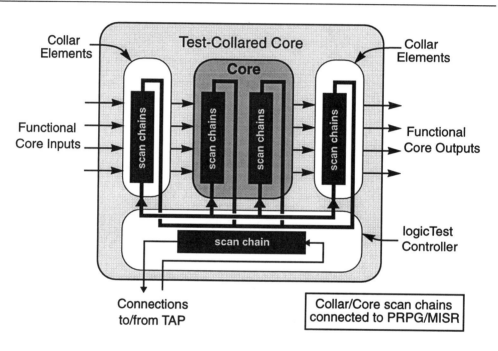

External Test Configuration

In the External Test configuration, the collar-element scan chains and scan chain of the logicBIST controller are connected to the chip-level test infrastructure. Figure 5-7 shows the External Test configuration. This configuration is used for testing surrounding UDL or interconnect between cores.

In the External Test configuration, the embedded core is completely isolated from the host chip by the collar elements. Therefore, the internal scan chains and logic of the embedded core do not have to be controlled. This allows the analysis at the chip level to proceed with a black box of the embedded core, which usually significantly reduces tool runtimes at chip-level integration. Also, because the chip-level analysis is independent of the embedded core contents, any changes that might be necessary within the embedded core do not require regeneration of chip-level test patterns.

Figure 5-7 External Test Configuration

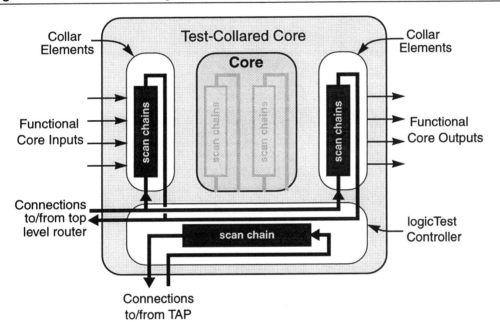

Direct Pin Internal Test

In the Direct Pin Internal Test configuration, the embedded core terminals are made accessible from chip pins. The test-collared core provides alternate inputs for each embedded core terminal input and alternate outputs for each embedded core terminal output. A multiplexer select signal allows the alternate inputs to control the embedded core inputs; alternate outputs are buffered versions of the embedded core outputs. The alternate inputs and outputs are automatically routed to the top level of the chip hierarchy. There they are either multiplexed with functional chip pins or connected to dedicated test pins if available. The Direct Pin Internal Test configuration is typically used for applying functional patterns to an embedded core from chip pins. If the embedded core contains more terminals than available chip pins the alternate signals can be further multiplexed at the chip pin. Thus a single chip pin can observe multiple core terminals from the same embedded core or multiple terminals from different embedded cores. Pin-sharing of this sort usually implies applying the same pattern multiple times and observing subsets of outputs during each pass.

The connections to and from the chip-level pins to the embedded core are automatically performed. These connections are also structurally extracted and verified by a rules checker. A simulation test-bench and test pattern are automatically generated to configure the design into the desired test configuration; multiple configurations might be required for a given embedded core if pin-sharing is required.

Figure 5-8 Direct Pin Internal Test Configuration

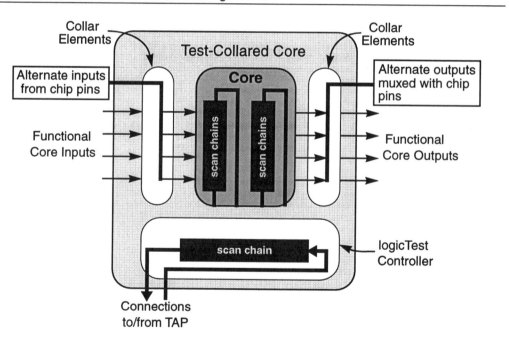

Test Collar Elements

This section describes in more detail the test collar elements used to access and isolate the terminals of the embedded core. A built-in set of collar elements is provided along with flexibility for the user to supply custom collar elements. These custom collar elements are automatically instantiated and connected similarly to the built-in cells. Elements are specified using a *Class* and *Subclass* field within a table file read by the automation tools. Table 5-1 lists the available classes of collar elements. Custom cells are also indicated using the class field.

Table 5-1 Collar Element Classes

Class Attribute	Description
I	Input
O	Output
IO	Bidirectional
<elementName>	Custom Element (either I, O, or IO)

Table 5-2 summarizes the subclasses. The subclass fully specifies the functionality of the cells used to construct a collar element.

Table 5-2 Collar Element Subclasses

Subclass Attribute	Functionality of the RTL Leaf Cell Associated with the Attribute
S	Inserts a scan flip-flop for observing and controlling an input or output port of the core.
P	Inserts the necessary logic to allow an input terminal of the core to be controllable from a chip pin or an output terminal of the core to be observable from chip pin. This functionality can be used to apply functional patterns to the embedded core from chip pins.
F0	Inserts the necessary logic to force the value of an input terminal of the core to a logic 0 during the *Internal Test* and *External Test* configurations. This functionality can be used to control asynchronous resets within the embedded core.
F1	Inserts the necessary logic to force the value of an input terminal of the core to a logic 1 during the *Internal Test* and *External Test* configurations. This functionality can be used to control asynchronous resets within the embedded core.
CD	Specifies capture-by-domain support for a collar element. An output collar element may require a capture disable signal during an *Internal Test* configuration if the embedded core output terminal data is sourced from a different clock domain than that of the collar element. An output collar element does not require a capture disable signal during the *External Test* configuration because the element holds its value during the capture cycle. An input collar element may require a capture disable signal during the *External Test* configuration if the input is receiving data from a different clock domain. An input collar element does not require a capture disable signal during an *Internal Test* configuration because the element holds its value during the capture cycle.

Composite Input Scannable Collar Element

Figure 5-9 illustrates an input collar element constructed with the S, CD, and P cell subclasses. For fully scannable cores, the P subclass is seldom required. The F0 and F1 cell subclasses, which are not shown in Figure 5-9, are typically used only for asynchronous inputs to the embedded cores.

Figure 5-9 Composite Input Scannable Collar Element

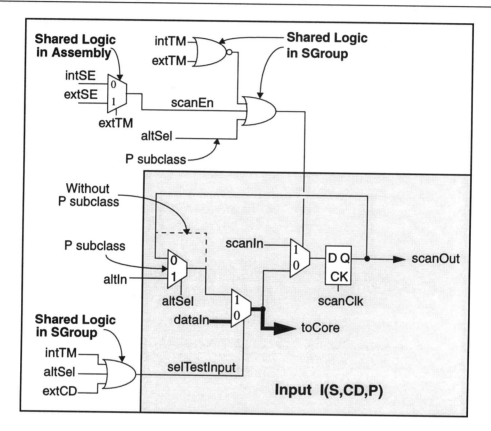

If the CD option is selected, the *extCD* signal blocks the capture of data from *dataIn*. This is required if the collar element receives data from a flop on a different clock domain. The *selTestInput* signal for input collar elements is driven by logic that can be shared by multiple cells. This logic also uses the *altSel* signal to ensure the *altIn* value is directed toward the core during direct chip pin test access mode. In internal test mode the input collar element holds its value during the capture cycle, so there is no need for additional logic. To prevent the flip-flop in the collar element from capturing data during the *Direct Pin Internal Test* configuration or during functional mode, the *scanEn* signal is held active. This is also done with shared logic.

Composite Output Scannable Collar Element

Figure 5-10 illustrates an output collar element constructed with the S, CD, and P cell subclasses. For fully scannable cores, the P subclass is seldom required.

Figure 5-10 Composite Output Scannable Collar Element

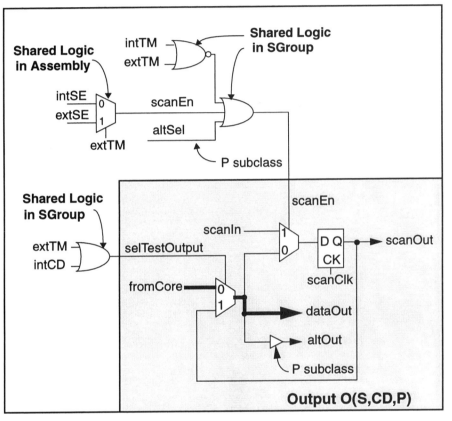

If the CD option is selected, the *intCD* signal blocks the capture of data from *fromCore*. This is required if the collar element receives data from a flop on a different clock domain. The *selTestOutput* signal for output collar elements is driven by logic that can be shared by multiple cells. In external test mode the output collar element holds its value during the capture cycle, so there is no need for additional logic. To avoid the collar element flop from capturing data during the *Direct Pin Internal Test* configuration or during functional mode, the *scanEn* signal is held active. This is also done by shared logic.

Design Flow

This section describes the design flow for embedded core test. There are two design flows for applying embedded logic test:

Core Provider/Chip Integrator Model

In this model, two parties are involved: the core provider and the chip integrator.

1. The core provider creates the test collar for a core.

2. The core provider makes the test-collared core available to the chip integrator, often as part of a library of cores from which the chip integrator chooses.

3. The chip integrator merges the test-collared cores into the design.

4. The chip integrator verifies the functionality of the embedded test infrastructure.

Chip Designer Model

In this model, the chip designer performs the following tasks:

1. Partitions a chip into several sections.

2. Treats some or all of these partitioned sections as cores.

3. Builds test collars for the cores.

4. Merges the test-collared cores into the design.

5. Verifies the functionality of the embedded test infrastructure.

The chip designer model is suited to designs whose complexity challenges the capacity of EDA tools. By using a divide-and-conquer approach, potential problems associated with tool memory requirements and tool runtime are eliminated.

Given that more complex system chip designs will contain third party IP or legacy cores, a test solution is needed that both facilitates testing the embedded cores and provides seamless integration with the rest of the design and test flow.

Additionally, as SOC design sizes continue to escalate, design partitioning becomes necessary because of EDA tool limitations and physical constraints such as power distribution. The divide-and-conquer approach of providing embedded test within a core simplifies the overall chip test problem.

The chip test problem, then, becomes a two-step process:

1. Each embedded core can be tested individually by an instruction or set of instructions from the TAP. For test non-mergeable cores, the core is isolated from its surrounding logic during the execution of its internal test. After the core is isolated, three test approaches are accessible through the TAP:

    •   Enable an embedded logicBIST controller that tests the core contents. This approach employs the *logic BIST* configuration.

    •   Apply functional patterns from external or package pins that are routed to the core. This approach uses the *Direct Pin Internal Test* configuration and is generally used for legacy cores.

    •   Apply scan patterns through the TAP or from package pins to scan chains of the core. This approach use the *Single-Chain Internal Test* configuration or *Multi-Chain Internal Test* configuration.

2. After each embedded core is tested, all interconnections between cores, or between cores and UDL, can be tested. Either a chip-level logicBIST controller can be used for an at-speed test or ATPG scan patterns can be externally applied for a slow-speed structural test.

Core-Level Design Flow

Figure 5-11 illustrates the complete design flow for the Core Provider/ Chip Integrator model. This flow describes the tool sequence for preparing an embedded core for reuse. The section "Chip-Level Design Flow" on page 135 describes the chip integration of such a reusable core. This flow assumes that the embedded core contains a legacy core and embedded memories.

Figure 5-11 Core Provider Design Flow

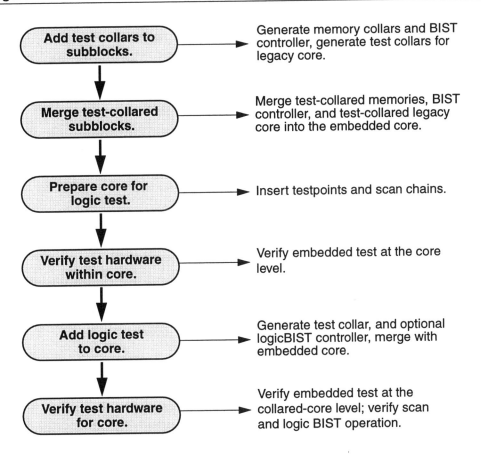

Chip-Level Design Flow

Figure 5-12 illustrates in more detail the chip-level integration of the test-collared embedded core. This flow describes the steps of integrating a reusable core that has been fully prepared and contains embedded logic BIST, a legacy core, and embedded memories. The integration of this complex embedded core into the host design follows a similar flow to that of its preparation; the collared core is integrated into the design immediately before testpoint and scan-chain insertion of UDL. Verification steps then follow, as shown in the figure.

Figure 5-12 Chip-Level Design Flow

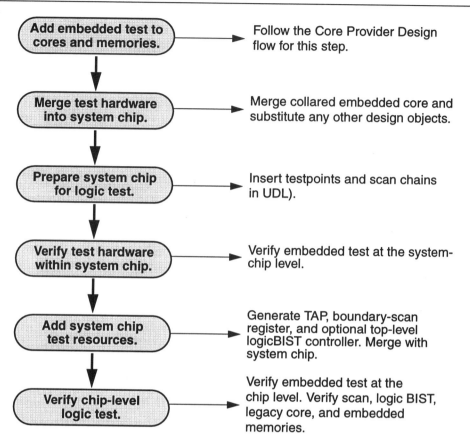

Add embedded test to
cores and memories.
→ Follow the Core Provider Design
flow for this step.

Merge test hardware
into system chip.
→ Merge collared embedded core and
substitute any other design objects.

Prepare system chip
for logic test.
→ Insert testpoints and scan chains
in UDL).

Verify test hardware
within system chip.
→ Verify embedded test at the system-
chip level.

Add system chip
test resources.
→ Generate TAP, boundary-scan
register, and optional top-level
logicBIST controller. Merge with
system chip.

Verify chip-level
logic test.
→ Verify embedded test at the
chip level. Verify scan, logic BIST,
legacy core, and embedded
memories.

Summary

Given the growing complexity of SOC designs, and the availability of VDSM process technologies, it has become necessary to evolve existing design methodologies. It is widely believed that a core-based design approach can significantly improve designer productivity. A corresponding test methodology that facilitates plug-and-play core re-use is essential for meeting aggressive time-to-market goals. The hierarchical core test solution described in this chapter provides a scalable architecture that supports complex SOC designs and very large ASIC designs. Flexibility is provided in supporting a variety of test techniques, including functional patterns, scan ATPG, and BIST. A two- step test process has been described that involves core preparation followed by chip integration. Design flows consisting of either the Core Provider/Chip Integrator model or the Chip Designer model are equally supported. Embedded test is an important part in this solution because of its inherent bandwidth scalability, ability to test at speed, and suitability for reuse at the board and system levels. Automation ensures repeatability and predictability of the process to meet time-to-market needs.

References

For a reasonably comprehensive bibliography of core-test-related papers, see the Web site http://grouper.ieee.org/groups/1500/bib.

[1] R. K. Gupta and Y. Zorian, "Introducing Core-Based System Design," *IEEE Design & Test of Computers*, 14(4):15–25, December 1997.

[2] Y. Zorian, E. J. Marinissen, and S. Dey, "Testing Embedded Core-Based System Chips," *IEEE Computer*, 32(6):52–60, June 1999.

[3] E. J. Marinissen and Y. Zorian, "Challenges in Testing Core-Based System ICs," *IEEE Communications Magazine*, 37(6):104–109, June 1999.

[4] Y. Zorian, "Test Requirements for Embedded Core-Based Systems and IEEE P1500," in *Proceedings of the IEEE International Test Conference (ITC)*, pp. 191–199, Washington, DC, November 1997, IEEE Computer Society Press.

[5] Y. Zorian, E.J. Marinissen, and S. Dey, "Testing Embedded Core-Based System Chips," in *Proceedings of the IEEE International Test Conference (ITC)*, pp. 130–143, Washington, DC, October 1998, IEEE Computer Society Press.

[6] IEEE P1500 Web Site. http://grouper.ieee.org/groups/1500/.

[7] IEEE Computer Society. IEEE Standard Test Access Port and Boundary-Scan Architecture – IEEE Standard 1149.1–1990. IEEE, New York, 1990.

[8] P. Varma and S. Bhatia. "A Structured Test Re-Use Methodology for Core-Based System Chips," in *Proceedings of the IEEE International Test Conference (ITC)*, pp. 294–302. Washington, DC, October 1998. IEEE Computer Society Press.

[9] E. J. Marinissen, R. Arendsen, G. Bos, H. Dingemanse, M. Lousberg, and C. Wouters. "A Structured and Scalable Mechanism for Test Access to Embedded Reusable Cores," in *Proceedings of the IEEE International Test Conference (ITC)*, pp. 284–293, Washington, DC, October 1998. IEEE Computer Society Press.

[10] L. Whetsel. "Addressable Test Ports: An Approach to Testing Embedded Cores," in *Proceedings of the IEEE International Test Conference (ITC)*, Atlantic City, NJ, September 1999. IEEE Computer Society Press.

[11] R. Kapur et al. "P1500-CTL: Towards a Standard Core Test Language," in *Proceedings of the IEEE VLSI Test Symposium (VTS)*, pp. 489–490, Dana Point, CA, April 1999. IEEE Computer Society Press.

[12] T. Taylor and G. Maston. "Standard Test Interface Language (STIL): A New Language for Patterns and Waveforms," in *Proceedings of the IEEE International Test Conference (ITC)*, pp. 565–570, Washington, DC, November 1996. IEEE Computer Society Press.

[13] T. Taylor, "Standard Test Interface Language (STIL): Extending the Standard," in *Proceedings of the IEEE International Test Conference (ITC)*, pp. 962–970, Washington, DC, October 1998, IEEE Computer Society Press.

[14] R. Kapur, E. J. Marinissen, T. Taylor, L. Whetsel, and Y. Zorian. Manuscript NL-MS 20.274 in preparation for the IEEE International Test Conference (ITC), Atlantic City, NJ, September 28–30, 1999.

Chapter 6

Test and Measurement for PLLs and ADCs

Embedded test can be used to implement efficient go/no-go or production tests and provide superior diagnostic capabilities. Embedded test can also be used to measure continuous variables of various kinds in an economical manner. This capability is necessary both for production testing and characterization of embedded components such as PLLs and ADCs. These components are accepted or rejected based on specifications that are expressed in terms of time, frequency, voltage, current, or unitless parameters.

This chapter describes the measurements performed by pllBIST (lock range, lock time, loop gain, and jitter) and adcBIST (offset, gain, second and third harmonic distortion, differential nonlinearity, and idle channel noise). The benefits of these solutions are explained as well as the design requirements to apply the methods.

Testing PLLs

pllBIST is a solution for at-speed testing of PLLs, and is delivered as software that generates synthesizable RTL code, a behavioral model of the PLL, synthesis scripts, testbenches, and a test pattern. The software facilitates automated insertion, verification, layout, and test generation.

The BIST circuit connects to only the input and outputs of the PLL to be tested; a digital multiplexer is inserted in the input path, but no changes or connections to the internal nodes of the PLL are needed. Operation is fully synchronous, making it suitable for high-speed operation.

The BIST circuit tests PLLs that contain edge-sensitive phase-frequency detectors (PFDs) and a charge-pump or other type of low-pass filter. It can accommodate PLLs whose feedback path includes a clock tree.

Various trade-offs between gate count, test time, and test accuracy can be made by setting appropriate parameters in the configuration file before generating the BIST circuit block.

PLL Concepts and Terminology

Because this book is primarily about digital design and test, some of the concepts and terminology pertinent to PLLs might not be familiar to all readers. This section briefly defines terminology relevant to testing PLLs. Complete definitions and descriptions are provided in References [1], [2], and [3] in the section "References" on page 169.

PLLs are used in many applications in ICs, such as the following:

● **Frequency synthesis.** The PLL generates a clock signal that has a frequency that is a constant multiple of a Reference Clock frequency and a phase that is constant relative to the Reference Clock.

● **Phase alignment.** The PLL generates a clock signal whose phase is aligned to some signal which may be advanced or delayed relative to the Reference Clock.

● **Clock generation.** The PLL generates a constant frequency clock whose phase is aligned to the average phase of a noisy Reference Clock which might have been derived from data.

Figure 6-1 Typical PLL Structure

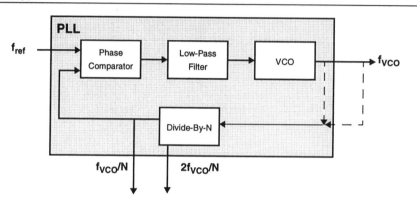

Figure 6-1 shows the structure of a typical PLL. The following are its primary components:

● A phase detector (PD) that compares the phase of the Reference Clock input to the output generated by the PLL and generates a digital pulse whose width is proportional to the phase difference.

● A low pass filter or integrator that generates an analog voltage across a capacitor; this voltage is made proportional to the average output voltage from the PD by converting the output voltage of the PD to a charging current via a resistor or charge pump.

● A voltage-controlled oscillator (VCO) that generates a digital signal whose frequency is proportional to the voltage at the input to the VCO; the output has an approximate duty cycle that is typically 50 percent.

● A divide-by-N (typically a simple binary counter) that divides the VCO output frequency down to the same frequency as the Reference Clock; this also delivers a signal with an exact duty cycle that is safer to use for clocking the digital circuitry of an IC.

Reference Frequency

This is the input frequency to which the PLL is trying to phase-lock its output frequency.

PFD

Phase Frequency Detectors are edge-sensitive circuits that have two outputs and two inputs. The inputs are a Reference Clock, which is the input to the PLL, and a Feedback Clock, which is an output of the PLL. The PFD is only sensitive to either the rising or the falling edges of the input signals, so the inputs can have any duty cycle. The outputs are Up and Down. Up is logic 1 when the Reference Clock phase is advanced relative to the Feedback Clock, and high impedance otherwise. Down is logic 1 when the phase is delayed (or retarded), and high impedance otherwise.

When the two outputs are used to drive a low pass filter or integrator, the resulting voltage is linearly proportional to the phase over a 360-degree range, and is hence sensitive to the frequency as well as the phase difference, allowing the PLL to be used over a wide frequency range. The stable operating point occurs when both inputs have approximately 0 degrees phase difference, neither Up nor Down is active, and no corrective phase action is generated (or needed).

Phase Comparator

Phase comparators are level-sensitive circuits that have one output. The best known example is a simple exclusive-OR gate. The two inputs are a Reference Clock and a Feedback Clock. The output is high when the two inputs have the same logic level, and low when the two inputs have different levels. The circuit is very fast, but when the two outputs are used to drive a low pass filter or integrator, the resulting voltage is linearly proportional to the phase over only a 180-degree range. Beyond 180 degrees the phase is ambiguous. For this reason, phase comparators tend to be used in very high-frequency applications or where the input frequency is constant and known in advance and only phase alignment is needed. The stable operating point occurs when the two inputs are approximately 90 degrees out of phase, causing the output to have an average value mid-way between logic 0 and 1.

Charge-Pump

A charge-pump is a circuit that linearly converts the duration of a digital pulse into a quantity of charge. It comprises a current source and a current sink (with nominally equal values). They are connected to an integrator or low pass filter via two switches that are controlled by the Up and Down outputs from a PFD. An integrator has infinite gain at DC (0 Hz), whereas a low pass filter has a finite gain (for example, gain=1).

VCO

A voltage-controlled oscillator generates a digital clock signal whose frequency is proportional to the voltage at the input to the VCO. The signal has an approximate duty cycle that is typically 50 percent. To obtain a signal with a precise duty cycle, the signal is divided by 2 or more using a binary counter. The most common type of VCO is the *current-starved* ring oscillator. It comprises an odd number of logic inverters connected in a ring, with the current available (from the power supply) to each inverter limited by the control voltage of the VCO. When the control voltage decreases the current flowing, the oscillation frequency decreases.

Phase Lock

Phase lock is achieved when the output signal no longer slips cycles relative to the input signal and the phase error becomes constant with some amount of random variation about its mean value.

Jitter

Jitter is the instantaneous time difference between the rising or falling edge of a signal-of-interest and the corresponding edge of a conceptual signal whose frequency is equal to the average frequency of the signal-of-interest. The statistical properties of this time difference are of more interest than the value for any one edge. The jitter of real signals (as opposed to simulated, noise-free signals) can be reported the same way that noise is reported: peak-to-peak, root mean square (RMS), or average (phase offset). See the "Noise" section on page 142 for definitions of these terms.

The value that is measured depends on the time interval in which the measurement is made, and how frequently the jitter value is sampled within that interval. The jitter can have a low frequency variation caused, for example, by 60 Hz on the power supply voltage. Provided that the measurement interval is much less than or much longer than the period of interfering frequencies, the measurement can be meaningful.

Interference that increases jitter can be caused by switching noise on the power rails. Switching noise refers to voltage spikes on V_{DD} or V_{SS} caused by logic gates switching. When a logic gate switches, it instantaneously requires current from the power supply distribution grid on the IC. Because the grid has nonzero resistance and inductance, a voltage drop can occur for the power rail in the immediate vicinity of the logic gate.

Loop Gain

This refers to the open loop gain, and is calculated by multiplying together the gains of every block in the PLL loop. The gain of the phase comparator is K_P, and equal to $i_{CP}/2\pi$ for a PFD with a charge-pump that sources/sinks current i_{CP}. The gain of the low pass filter or integrator is a complex function of frequency. The gain of the VCO is K_V, in units of radians/sec per volt, for example. The divide-by-N has a gain of $1/N$.

Lock Range

This is equal to the maximum lock frequency minus the minimum lock frequency. Usually the specific maximum and minimum are of interest rather than their difference. The maximum lock frequency is conventionally measured by first achieving phase-lock at some mid-range frequency, and then slowly increasing the input frequency until phase-lock is lost (because the loop gain decreases too much). The maximum rate at which the frequency is increased must be specified.

Noise

Noise, in the context of electronic circuits, refers to random (or apparently random) variations in the electrical value of a parameter, as seen in the top graph of Figure 6-2 on page 143. If the random variable is sampled many times, and the probability of each value is plotted against the value, naturally occurring random values tend to have bell-shaped curves, as seen in the bottom graph of Figure 6-2. This is referred to as a Gaussian or "Normal" probability distribution function (PDF), and it has very predictable mathematical properties. The mean value and the standard deviation summarize the whole distribution. Another property of interest is the peak-to-peak value, which is infinite in theory, but, in practice, is typically 6 to 14 times the standard deviation.

For any distribution, the mean value is the average value. The standard deviation is denoted by the Greek letter sigma (σ), and is also known as the root mean square (RMS) value, when the mean is zero, because it is equal to the square root of the mean of the squares of the values. For a Normal distribution, 68.3 percent of the values lie within one sigma of the mean. One sigma indicates the width of the central portion of the bell-shaped Normal curve, as measured between the points of curve inflection. The peak-to peak value is the maximum recorded value minus the minimum recorded value, and all samples lie within this range.

The cumulative distribution function (CDF) is the integral of the PDF. It shows that 84.1 percent of the values are less than +1 sigma, and 15.9 percent of the values are less than -1 sigma.

Figure 6-2 Normal (Gaussian Noise) Distribution and Its Statistical Summary

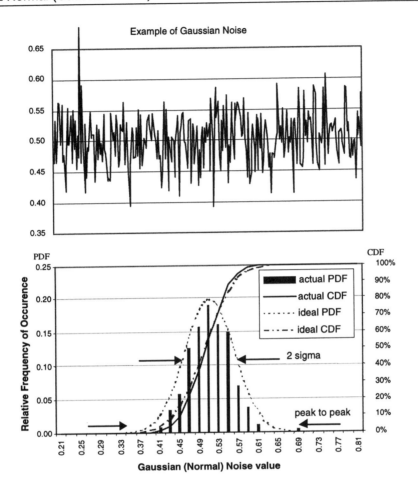

Function Summary

pllBIST uses the externally supplied reference clock input to the PLL as a constant frequency stimulus, and generates an on-chip variable frequency stimulus for connection to the PLL input via an on-chip multiplexer. Clock outputs of the PLL, and a lock indicator if used, are processed by the BIST circuit to calculate loop gain, maximum and minimum lock frequencies, lock time, and RMS or peak-to-peak jitter.

Binary-encoded upper and lower limits for every measurement are serially shifted into the IC via an 1149.1 TAP controller or other test access circuit. The measurement values are serially shifted out via the TAP, along with a pass/fail result for each limit comparison. This allows the automatically generated test pattern to comprise only logic 1 and logic 0 values with no tester-specific programming needed.

Test times can be increased (or decreased) relative to the times shown in Table 6-1 on page 152, if more (or less) accuracy is needed.

Measurement Summary

pllBIST performs tests to measure six important specifications (loop gain, maximum and minimum lock frequencies, lock time, RMS and peak-to-peak jitter), typically using the circuit illustrated in Figure 6-3. The tests performed are jitter, lock range, lock time, and loop gain.

Multiple PLLs can be tested with a single BIST controller by adding a multiplexer at each connection between the controller and the PLL, and selecting the chosen input to the multiplexers via userBits provided within the controller, within the TAP controller, or elsewhere. However, the loop gain test accuracy decreases with increasing path delay between the PLL and the BIST circuitry, so loop gain should be measured for only one PLL which is in close proximity. Fortunately, PLLs of similar design on the same chip are likely to have well-correlated loop gains, and localized faults will likely be detected by the other pllBIST tests.

Figure 6-3 Typical Connections Between BIST and a PLL

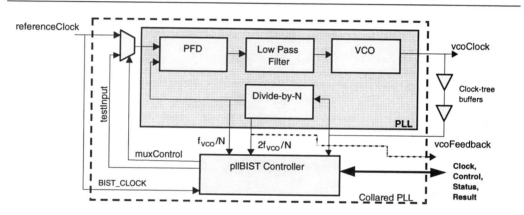

Primary Benefits

The following are the primary benefits of using BIST for PLLs:

- Automated generation of test patterns suitable for any tester
- Automated generation, insertion, and verification of the test circuitry
- Thorough testing of PLLs using low-cost, digital ATE
- Faster testing of a PLL than is typically achieved with external ATE
- Testing at input frequencies higher than the ATE capability
- Reuse of the PLL test at the IC, board, and system levels

Measurements for PLLs

Each of the tests described in this section is individually enabled via the test access port, which is typically the 1149.1 TAP. For a more technical description of the tests (patent pending), and results from real chips, see Reference [4] in the section "References" on page 169.

Jitter

The Jitter test measures jitter over a user-specified number of input clock cycles. In effect, a histogram of the jitter is calculated and the user can select where to define the limits: peak-to-peak or +/- 1 sigma (to obtain the RMS value).

While a jitter-free clock is applied to the PLL, BIST measures the jitter in the PLL's digital output over a prescribed number of clock cycles.

It is assumed that a sufficiently jitter-free clock, at the desired test frequency, is available from an off-chip source and is applied to the BIST_CLOCK input of the BIST controller. The referenceInput signal can have any amount of jitter, and can even contain serial data, because the signal is only used to drive the PLL to the output frequency at which measurements will be initiated.

As Figure 6-4 illustrates, the BISTclock signal (which could also be the *referenceClock* input to the PLL) clocks a rising-edge triggered, D-type flip-flop after being delayed by a digitally programmable delay line. An output from the PLL is the data to the flip-flop after being delayed.

Figure 6-4 pllBIST's Jitter Measurement Circuitry

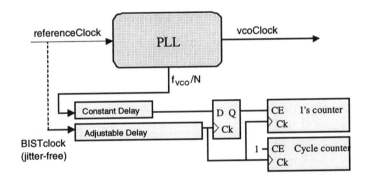

By recording the number of logic 1's output from the flip-flop in Figure 6-4, and digitally varying the delay through the adjustable delay, a cumulative distribution function (CDF) of the output digital codes is produced, usually similar to the one shown in Figure 6-5 (see also Figure 6-2 on page 143). For example, for the delay control setting b_100000, the number of times that a logic 1 is output might be 127 times out of 256, corresponding to 50 percent. For the delay setting b_100001, the number might be 191, corresponding to 75 percent.

Figure 6-5 CDF of D-Type Flip-Flop Output: Percentage of 1 Values Versus Variable Delay Setting

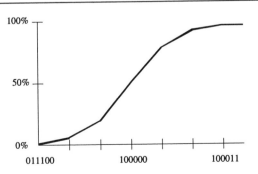

The difference between the delays corresponding to 0.1 percent and 99.9 percent is approximately equal to the peak-to-peak jitter. The difference between the delays corresponding to 16 percent and 84 percent is approximately equal to twice the RMS jitter. Using a ring oscillator that includes the adjustable delay line, the oscillation period is measured for each delay setting (for example, the delay setting for which 16 percent of the flip-flop's output were logic 1 values, and the delay setting for 84 percent) and one period is subtracted from the other. This gives a binary-coded, user-defined jitter measurement in units of time (for example, picoseconds), independent of the logic gate delays used in the variable delay line.

Immediately before running the test, the lower percentage limit is shifted in, along with the number of clock cycles for which jitter is to be measured. The upper percentage limit is automatically 100 percent minus the lower limit. After the test is completed, the binary-encoded upper and lower limits for jitter are serially shifted into the IC, via the TAP. Then, the value of jitter is shifted out twice, each time followed by a pass/fail bit for a limit.

The tester needs only to check each pass/fail bit to determine whether the PLL passes each test. The measured value can be stored in the tester for later statistical analysis if desired. The shifted-in limits can be any value if the pass/fail bit is to be ignored, for example, during characterization.

Optionally, this test can continue into a characterization mode in which the complete CDF is measured. In this case, the percentage of logic 1 values detected for each setting of the adjustable delay is output, in sequence, one per shift-out. Next, the oscillation period for each setting of the variable delay is output, in sequence, one per shift-out. Using all these binary-encoded values, a complete CDF plot can be generated, which is useful for ensuring that the distribution is Normal, and for seeing whether the shape changes for different integration intervals (number of clock cycles) which might indicate a frequency component in the jitter.

To ensure that jitter in the BIST circuitry itself does not cause inaccurate measurements, a mode is provided for measuring self jitter. In this mode, the jitter in one of the two delay lines in the BIST circuitry is measured relative to the other delay line. Jitter in either delay line will affect the measurement result, if the jitter in each delay line is uncorrelated to that in the other. The primary purpose of this test is to detect manufacturing flaws or a layout with insufficient power rail stability. Pass/fail limits and measurement results are shifted in and out in the same fashion as when measuring jitter in the PLL.

pllBIST can measure cycle-to-cycle jitter, that is, the instantaneous variation in period from one period to the next in the output signal from the PLL. In this mode, jitter in the PLL output signal is measured relative to itself after travelling through a delay line whose delay is digitally varied between the clock period plus 2 nanoseconds, say, and the clock period minus 2 nanoseconds. Cycle-to-cycle jitter tends to be much less than long-term jitter, and might not reflect accumulated large excursions from the mean frequency. However, the BIST circuit samples every output cycle instead of every N^{th} cycle (see the section "Accuracy Limits for Jitter Test" on page 147), so the measurement can be more revealing. In this test mode, at extremes in temperature/voltage/process, the BIST circuit might not be able to deduce where the next cycle begins because of the large change in the nominal delay of the delay lines. For this reason, this test is best performed at mid-range input frequencies or at nominal conditions. A key advantage of this test is that the PLL's input frequency is not used as a jitter-free reference.

Accuracy Limits for Jitter Test

The accuracy of the jitter measurement is limited by several factors:

- Digital resolution (delay quantization) of the variable delay line; that is, the delay of a logic gate

  Faster technologies permit finer resolution. pllBIST uses a proprietary digital technique to achieve time resolution equal to approximately one-eighth of an inverter delay.

- Jitter measurement interval may be too short

  For noisy phenomena, theory predicts that the true standard deviation will be within 5 percent of the estimate (99-percent confidence level) if more than a thousand samples are used for the measurement. Also, the mean phase error can change over time (especially within time intervals less than the period of the PLL's natural frequency). Because the 16 percent and 84 percent points of the CDF are measured in different intervals, a change in the mean phase error between each of these two intervals will introduce error.

- Jitter sampling on every edge of the *referenceClock*; that is, every N^{th} edge of the VCO output clock

  N-1 out of every N clock edges of the VCO output clock are not sampled. The unsampled edges could contain more jitter than the sampled edges, and this jitter would not be measured.

- Jitter in the PLL caused by BIST

  Good isolation between PLL and BIST circuit power rails is important, and is best achieved by using power supply pins for the PLL that are different than those used for BIST and other logic. This is standard layout practice for PLLs and other analog circuitry.

- Jitter in the variable delay line within the BIST circuitry

  Good power supply quality (as seen by BIST) is important and is achieved in IC layouts that minimize wire inductance and resistance. This type of jitter will be present in any digital path in the IC and so will similarly affect jitter measured by external testers.

- Jitter in BIST_CLOCK

  The jitter is measured relative to this clock, so any (uncorrelated) jitter in this clock will tend to be added to the value measured. The addition will be linear for the peak-to-peak value, and nonlinear for the RMS value (it will be the square root of the sum of the squares). This source of inaccuracy will similarly affect jitter measured by external testers.

Values for *m* (Jitter Measurement Interval)

There will be variation (or, more mathematically, variance) in the measurement result because of random noise, system noise, and quantization. To improve test repeatability, which allows smaller guardbands or higher yield, the variance can be reduced by increasing the value of *m* (the jitter measurement interval, in cycles of f_{REF}). However, taking more samples increases test time, as seen in Table 6-1 on page 152. Generally, *m* is determined by system requirements.

Jitter typically has both high and low frequency components. Input jitter at frequencies below the natural frequency (f_N) of the PLL tends to be tracked by the PLL and does not cause jitter at the output; input jitter at higher frequencies gets filtered out by the PLL. To measure only the higher frequency jitter at the output of the PLL, the jitter must be sampled for a shorter duration (smaller *m*). However, the mean phase error can change between sample intervals for intervals less than the period of the natural frequency, which can introduce error. As a guideline, *m* should be at least 10 times the ratio of f_{REF} to f_N.

Measuring jitter over a longer interval (larger m) will include output jitter at both low and high frequencies. The PLL's natural frequency will typically be between 0.1 and 0.001 times the PLL's input frequency, so setting m to 8192 or larger will ensure that all frequency components of interest are included. Running the BIST jitter test twice, once with $m=8191$ and once with $m=127$, can be useful for checking the frequency dependence of jitter. The variance of the result when $m=127$ will be relatively large because fewer samples are averaged.

Lock Range

The Lock Range test measures the minimum or maximum (selected at test time) output frequency of the VCO when it is controlled via the PLL's phase comparator input. These frequencies correspond, in most cases, to the lock range of a PLL with a PFD.

pllBIST applies a stimulus frequency to the PLL which is always a little higher (or lower) than the PLL's present output frequency. The PLL output will smoothly ramp to its maximum (or minimum, respectively) output frequency, which is then measured. This frequency corresponds to the upper (or lower) extent of the frequency lock range.

The PLL is assumed to have a VCO that has a nonzero minimum output frequency and that saturates at some maximum frequency.

Using the PLL VCO output, signal $f_{TestInput}$ is generated at half (or twice) the PLL's internal feedback frequency, and *TestInput* is selected via the multiplexer at the input to the PLL. The PLL output frequency ramps down (or up, respectively) linearly and continuously until it reaches some lower (or upper) limit. By continuously measuring the PLL's output frequency, the BIST circuit determines when the output frequency approaches its minimum (or maximum) value.

Next, the binary-encoded, upper and lower limits for minimum (or maximum) lock frequency are serially shifted into the IC, via the TAP. Then, the binary-encoded value of the lock range frequency is shifted out twice, each time followed by a pass/fail bit for a limit.

The preceding test can be run twice, once for the minimum frequency of the lock range, and once for the maximum frequency.

If the VCO is designed using a *current-starved* ring oscillator, its minimum oscillation frequency might be zero, that is, when the control voltage is less than the threshold of an *n*-channel transistor, the current may drop to zero. This is difficult to test with BIST, and can also cause start-up problems in normal function-mode operation. Many current-starved ring oscillator designs now include an internal weak pull-up resistor such that a zero volt input results in some nonzero oscillation frequency. BIST might measure this frequency as the minimum lock frequency, though it is possible that the loop gain at this frequency might be too low to achieve phase locking. A practical method to avoid the zero frequency problem, is to provide the minimum frequency as f_{REF}, and if the PLL locks, its operation at that frequency is verified; BIST can then automatically test for just the maximum frequency.

By reducing the frequency measurement interval, it is possible to determine the frequency at which the loop gain diminishes significantly (the frequency versus voltage curve flattens) which may give a better estimate of the lock range.

Accuracy Limits for Lock Range Test

The accuracy of the lock range measurement is limited by two factors:

- Non-saturating frequency versus voltage response for the VCO— BIST will force the control voltage of the VCO to go to its maximum and (or minimum) value, whereupon the frequency is measured. If this frequency is not representative, then the measurement might be invalid. As discussed previously, this can occur for a current-starved ring oscillator. Another possibility is a VCO which shows a peak in its transfer function, or a VCO that has a maximum frequency higher than the digital logic can handle.

- The measurement interval programmed at test time—too large a value will result in a value that corresponds to the VCO output frequency at its saturated value, at which loop gain is close to zero and phase-lock is tenuous. Too small a value will have too much variance.

Lock Time

The Lock Time test measures the lock time for the PLL, when the input frequency makes a step change from the maximum or minimum (selected at test time) to the nominal frequency. This is usually a worst-case frequency deviation and lock time.

The PLL output frequency is forced to its maximum or minimum, the nominal input frequency is reapplied, and the number of clock cycles is counted until lock is regained.

pllBIST feeds back an appropriate signal to the input of the PLL to force the PLL to go to its maximum or minimum output frequency—as selected by a bit scanned in during invocation of this test.

Next, the nominal input signal is selected via the multiplexer at the input to the PLL, and counting of timebase clock cycles begins. Another of the scanned-in bits selects which lock indicator is to be used: a (phase) lock indicator provided by the user, or the frequency lock detector provided within the BIST circuit. Frequency lock is indicated when the output of a frequency counter, driven by the PLL's output frequency, is within +/-1 count of the correct frequency count (scanned in when the test is invoked).

Next, the binary-encoded, upper and lower limits for lock time are serially shifted into the IC, via the TAP. Then, the binary-encoded value of the lock time is shifted out twice, each time followed by a pass/fail bit for a limit.

Accuracy Limits for Lock Time Test

The accuracy of the lock time measurement is limited by the following factors:

- The difference between the maximum/minimum frequency and f_{REF}—if this is large relative to the typical frequency change, then the lock time measured may be dominated by the time to charge the capacitor in the loop filter, and will be longer than is typical. The test performed by the BIST measures the *maximum* lock time.

- Incorrect PLL operation when the VCO control voltage is at its maximum or minimum value— this is the starting point for the lock time interval, so if such a maximum/minimum voltage results in non-oscillation, for example, then the lock time test will be inaccurate.

Loop Gain

The Loop Gain test measures loop gain for a positive or a negative (selected at test time) step change in input phase.

pllBIST applies a step change in phase to the PLL, and measures the resulting change in the PLL's output frequency.

The PLL is first driven by its normal input, f_{REF}.

A frequency measurement unit is included in the BIST circuitry to count the number of PLL output cycles in a constant number of timebase clock cycles. Typically, the timebase clock frequency f_{BIST_CLOCK} is equal to f_{REF} (that is, is connected to *referenceClock*).

Using the PLL VCO output, the BIST circuit generates a signal, *TestInput*, which initially has the same phase and frequency as f_{REF}. After the PLL has locked to f_{REF}, *TestInput* is selected via the multiplexer at the input to the PLL. This action effectively opens the loop of the PLL. Next, a phase delay (or advance) is generated for one or more cycles of $f_{TestInput}$. The number of cycles modified is denoted as *NumberOfModifiedCycles*. During these phase shifts, the charge-pump will only sink (or source, respectively) charge from (or into) the integrator. After the modified cycles, the phase change returns to zero, at which time the PLL's new output frequency is measured (a calibration routine is included in this step to account for any residual phase error or charge-pump leakage). The change in frequency, relative to f_{REF}, is proportional to the open loop gain, G_{OL}, of the PLL, according to the following equation:

$$G_{OL} = \frac{K_V i_{CP}}{4\pi NCf_{REF}^2},$$

where K_V is the gain of the VCO (radians/sec/volt), i_{CP} is the charge- pump current (sink or source, as appropriate), N is the value from the digital divider, C is the integration capacitance, and f_{REF} is the input reference frequency.

This open loop gain is proportional to the square root of the PLL's natural frequency as follows:

$$G_{OL} \cong \left(\frac{2\pi f_N}{f_{REF}}\right)^2,$$ where f_N is the natural frequency in Hz.

Next, the binary-encoded, upper and lower limits for loop gain are serially shifted into the IC, via the TAP. Then, the binary-encoded value of the loop gain is shifted out twice, each time followed by a pass/fail bit for a limit.

The preceding test should be run twice, once for a phase advance and once for a phase delay (in either order) during an equal number of modified cycles. Because the loop gain can be measured for both current sinking and sourcing into the integrator, any mismatch in these currents can be detected. However, if the PLL uses a voltage driver instead of a charge-pump, mismatch in loop gain measurements is also proportional to the difference between f_{REF} and the mid-range frequency (more specifically, it's proportional to the difference between the actual VCO control voltage and the control voltage's mid-range value). In this case, the sum of the unsigned loop gains for current sink and source should be tested since it is independent of f_{REF}, in the middle of the lock range.

The equation for loop gain that is presented here (above) is slightly different from most equations reported in the literature because there are many ways to measure the loop gain. The important consideration, from the standpoint of parametric fault coverage, is that the loop gain measured has a known relationship to the gain of all the important elements in the PLL's loop.

Accuracy Limits for Loop Gain Test

The accuracy of the loop gain measurement is limited by several factors:

- Delay of the multiplexer in the input path

  Shorter delays are preferred and can be achieved with the usual techniques, such as manual layout and faster technologies.

- Value of N in the PLL's divide-by-N

  Smaller values mean that the output phase is updated more frequently, so the "open loop" response is more sensitive to the multiplexer's delay.

- Value of *NumberOfModifiedCycles*

  Larger values can cause the integrator to saturate. To determine the point of saturation, this parameter can be increased until the value output by the BIST circuit starts to increase non-linearly.

- Difference in frequency between f_{REF} and the closest extreme of the Lock Range, when a low pass filter is used

  The middle of the range is best (this limitation does not apply for charge-pumps or integrators). Note that integrators have infinite gain for a constant phase offset input, whereas low pass filters have a finite gain.

- Linearity of the PLL's integrator

  This factor is not usually significant.

Test Times

The Loop Gain test requires the PLL to first attain frequency lock. First, the loop gain for a phase advance is measured (<1000 cycles of VCO output frequency), which causes the PLL output frequency to become different from the input frequency. Next, frequency lock is again achieved, and the loop gain for a phase delay is measured. Lock time for most PLLs is less than 1000 input clock cycles, so test time should be less than 3000 clock cycles (0.3 ms, at 10 MHz).

The Lock Range test requires the PLL output to smoothly ramp down to its lowest frequency, starting from its nominal frequency, and to then ramp up to its highest frequency. This will typically take the equivalent of two lock times, plus the time for two frequency measurements. In other words, test time is about the same time as the Loop Gain test (0.3 ms, at 10 MHz).

The Lock Time test requires one lock time, which is typically less than 1000 input cycles (0.1 ms, at 10 MHz). This lock time might be greater than the typical case in which a small frequency step is applied.

The Jitter test requires J*m input cycles (by default) to determine the two delay settings that correspond to the jitter limits. J is the number of delay settings searched for the adjustable delay line, and m is the user-specified number of clock cycles over which jitter is measured (and is rounded up to 2^K-1, where K is an integer). BIST circuit generation software calculates the maximum value for J by dividing the achievable delay increments into the maximum expected peak-to-peak jitter. It adds a "safety" factor of two, and assumes the expected phase error is zero. An additional 1024 input cycles are needed to convert these settings into a time difference. For a typical J=128, the test time is 128m+1024 input clock cycles. Using the default value of m=1023, the test time is 13.2 ms at 10 MHz.

In summary, for the sample 10-MHz PLL, test times are as follows:

Loop Gain	0.3	(loop gain, up and down)
Lock Range	0.3	(lock range, upper and lower)
Lock Time	0.1	(lock time)
Jitter	13.2	(RMS jitter, in 1024 cycles)
Total	14 ms	

Table 6-1 lists test times for a variety of other jitter measurement intervals and input clock frequencies.

Table 6-1 Test Times for Various Frequencies and Jitter Intervals

f_{REF}	m $(m = 2^K - 1)$	Loop Gain + Lock Range + Lock Time Tests 7000/f_{REF}	Jitter Test $(128m + 1K)/$
(MHz)	(input clock cycles)	(ms)	(ms)
10	127	0.70	2
10	1K	0.70	13
10	8K	0.70	106
50	1K	0.14	2
50	8K	0.14	22
50	64K	0.14	168
100	8K	0.07	10
100	64K	0.07	84
100	512K	0.07	672

Assumptions for Testing PLLs

To economically achieve the required measurement accuracy and resolution, several assumptions are made about the PLL to test and the environment in which it is located. If these assumptions are not already valid for an IC design, the following guidelines help make the assumptions valid.

The assumptions are as follows:

- The PLL has a digital input and outputs.

- A synthesized divide-by-N can be clocked at the same rate as the divide-by-N within the PLL. If the synthesized divide-by-N cannot be clocked at the same rate as the divide-by-N within the PLL, then the schematic synthesized from the RTL code will require manual layout.

- The jitter in the PLL is not increased by the switching noise of various logic (including delay chains) within the BIST circuitry. Separate power supply rails should be used for the BIST circuit and the PLL. Refer to the section "Noise" on page 142.

- The jitter on the output of a typical logic gate, relative to its input, is much less than the PLL's output jitter. A *clean* power rail is the primary way to ensure this state.

- Means are available for disconnecting the PLL's normal input signal, and for connecting the stimulus generated by the BIST controller to the PLL input, with less than 1-nanosecond propagation delay (for example, a multiplexer or tri-state buffers).

- The delay added to the PLL's input path by inserting the 2-to-1 multiplexer is not critical. If the delay is critical, then manual placement of the multiplexers can be done, or matching delays can be added elsewhere.

- A single PLL is tested. If more than one PLL is tested, a register and multiplexers, external to the BIST controller can be added to enable testing multiple PLLs.

- A jitter-free stimulus clock is available to drive the PLL within its frequency capture range.

- Testing of the BIST circuitry will be addressed by the BIST used to test other logic on the IC, or an ATPG scan path test from off chip.

Testing ADCs

adcBIST is a solution for at-speed testing of ADCs. It is delivered via software that generates synthesizable RTL code, a behavioral model of the ADC, synthesis scripts, testbenches, and a test pattern. The software facilitates automated insertion, verification, layout, and test generation.

The BIST circuit connects to only the input and outputs of the ADC to be tested, and no changes or connections to the internal nodes of the ADC are needed. Operation is fully synchronous with the sampling clock, making it suitable for low-or high-speed operation.

adcBIST can test almost any type of ADC, including flash, successive approximation, and sigma-delta. It can accommodate AC-coupled inputs, pipelining, and differential analog inputs.

Various trade-offs between area, test time, and test accuracy can be made by setting appropriate parameters in the configuration file before generating the BIST controller circuit block.

ADC Terminology

This section briefly defines terminology relevant to testing ADCs. Complete definitions and descriptions are given in References 5 and 6 in the section "References" on page 169.

ADC

ADCs are used in many applications in integrated circuits, such as the following:

- Telephony—the ADC converts voice signals into pulse code modulation (PCM) digital signals, at a rate of 56K or 64K samples per second

- Telecommunications—the ADC samples high-speed digital signals, whose waveform has been significantly altered by transmission on a telephone line or video cable, to allow digital signal processing, which reconstructs the original digital signals

- Computers—low-speed ADCs sample analog inputs from various sources, such as a microphone, sensor, or battery voltage

The structure of the ADC is typically one of the following:

- Flash converter

- Successive approximation register

- Pipelined converter

- Delta-sigma

Delta-Sigma (or Sigma-Delta) Converter

This type of ADC comprises two sections, a delta-sigma modulator and a decimator.

The modulator uses relatively few analog components. It continually compares the analog input signal to one of two DC voltages, integrates the error voltage produced from this comparison, then digitizes this integral into a single bit (or a small number of bits). The bit selects which of the two (or more) DC voltages is used for the initial comparison, so as to minimize the average error over time. The generated bit may also proceed to more stages of the modulator to reduce the error over different time intervals. The proportion of logic 1 values in the final output delta-sigma bit stream, in any time interval, represents the average value of the analog signal in that interval. The sample rate is typically 32 to 64 times the frequency of the highest frequency of interest in the analog signal.

The decimator is a digital circuit that continually averages bits in the delta-sigma bit stream to obtain multi-bit words at a sample rate which must be at least twice the frequency of the highest frequency of interest in the analog signal.

There will typically be several master clock cycles of delay through this type of converter, to the delta-sigma bit stream output (which may be accessed directly for testing gain and harmonic distortion), and more cycles through the digital decimator (which must be accessed for noise testing to ensure that the measured noise is suitably band-limited).

Flash Converter

The analog input signal is compared to 2^N-1 different reference voltages with 2^N-1 comparators, and the "thermometer scale" digital result is converted to a binary-encoded value with N bits. The sample rate must be at least twice the frequency of the highest frequency of interest in the analog signal, and an N-bit digital value is generated every cycle of the master clock whose frequency is equal to that of the sampling clock. There will typically be one master clock cycle delay through the converter.

Pipelined Converter

The analog signal is compared in the first stage, to a mid-rail voltage, and if it is greater than this voltage the most significant bit is a logic 1 (otherwise a 0); then the mid-rail voltage is subtracted from the input signal (otherwise no action), and the result doubled; in the next stage (which is the same design as the first) the output of the previous stage is compared to the mid-rail voltage, and so on; in this way, after N stages are performed in series, N binary-encoded bits are generated.

While the second stage is processing the result from the first stage, the first stage takes the next sample of the input analog signal. The sample rate must be at least twice the frequency of the highest frequency of interest in the analog signal, and an N-bit digital value is generated every cycle of the master clock whose frequency is equal to that of the sampling clock. There will typically be N master clock cycles of delay through the converter.

Successive Approximation Converter (SAC)

This type of converter typically comprises an array of capacitors with binary ratios. One terminal of each capacitor is connected to all the other capacitors and a comparator, and the other terminal is selectively switched. The common capacitor terminal voltage is compared to a mid-rail voltage, and if it is less than this voltage the most significant bit is a logic 0 (otherwise a 1). Each capacitor, starting from the most significant bit (MSB) capacitor, is switched in turn and the bit values determined. After N switchings are performed, N binary-encoded bits are generated. The sample rate must be at least twice the frequency of the highest frequency of interest in the analog signal, and an N-bit digital value is generated every N cycles of the master clock, whose frequency must be N times the sampling rate. There will typically be N master clock cycles of delay through the converter.

Offset

The offset, as measured by adcBIST, is the difference between the digital equivalent of the average analog voltage applied and the average output digital value. For a 3-bit ADC that converts voltages between 0 and 8 volts, if a linear ramp is applied which increases from 0 to 8 volts, and the average digital output is 100.0 (binary), the offset is a half LSB. The average output should be the average of 000, 001, 010, 011, 100, 101, 110, 111, which is 011.1 (binary).

Another definition for offset is the difference between the output and the exactly middle digital code, when an exactly mid-rail voltage is applied to the ADC input.

Gain

The gain, as measured by adcBIST, is the slope of the straight line that best fits the time-domain transfer function of the converter.

Another definition for gain is the slope of the line that joins the end-points of the time-domain transfer function.

Harmonic Distortion

When a pure sinusoid of frequency f_1 is applied to an ADC, the harmonic distortion at the ADC output refers to the level of the sine wave amplitude at multiples of f_1, which is ideally zero. In reality, there will at least be distortion due to quantization and because of imperfect binary ratios in the ADC. adcBIST reports harmonic distortion as the ratio between the amplitude of the specific harmonic and the amplitude of a maximum amplitude sinewave (that is, one that swings between the logic 0 and logic 1 voltages of the 3-state driver). If the ADC were perfectly linear, aside from quantization distortion, then zero distortion would be reported, whereas a Fourier transform would report the quantization distortion (= N*6 dB, where N is the number of bits).

Noise

Refer to the section "Noise" on page 142 for a definition.

Quantization

This term is used when converting a continuous variable, such as time or voltage, into a digital value. When converting a continuous variable with range R into a digital word with N bits, 2^N values are possible. The error introduced by quantization has a maximum value of $R/2^N$. This is also referred to as the resolution, or the value of one least significant bit (LSB).

Function Summary

adcBIST generates an analog stimulus for connection to the ADC via a multiplexer or 3-state driver, and a single resistor, R, and capacitor, C, (with RC = $R_{NOM}C_{NOM}$ +/-20 percent). The RC can be on chip or off chip, depending on the sample rate, number of bits, number of samples, RC linearity, and ADC input impedance. The nominal value of RC is calculated automatically by software. As the ADC samples the stimulus, its output is processed by BIST circuitry using a novel technique (see Reference 7) to calculate the offset, gain, second and third harmonic distortion. Other methods are used to measure the wideband RMS or peak-to-peak idle-channel noise, and differential non-linearity (DNL) at all or selected steps.

Binary-encoded upper and lower limits are serially shifted into the IC via an 1149.1 TAP controller or other test access circuit. The measurement values are serially shifted out via the TAP, along with a pass/fail result for each limit comparison. This allows the automatically generated test pattern to comprise only logic 1 and logic 0 values with no tester-specific programming needed.

Measurement Summary

adcBIST performs three tests to measure six standard specifications, typically using the circuit diagram illustrated in Figure 6-6. The tests are best-fit polynomial, DNL, and idle channel noise.

Figure 6-6 Typical Connections Between BIST and ADC

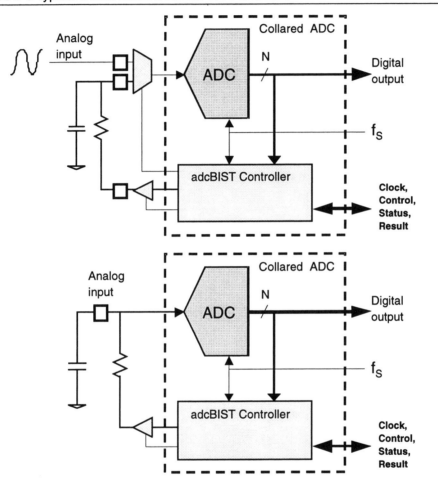

These tests support the following standard specifications for ADCs, which are defined in the section "ADC Terminology" on page 153.

- Best-Fit Polynomial
- Offset
- Gain
- Harmonic Distortion
- DNL
- Idle Channel Noise

Multiple ADCs can be tested with a single BIST controller by connecting additional 3-state drivers to the BIST stimulus output, each one enabled by a bit in the *userBits* register in the BIST controller, the TAP controller, or from elsewhere. A corresponding N-bit wide multiplexer is enabled by the *userBits* to sample the appropriate ADC's output. The ADCs may have different numbers of bits, as long as all most-significant bits (MSBs) are multiplexed into the MSB input of the BIST controller, and the BIST is compiled for the largest number of bits in any one ADC. The voltage range for each analog input may be different, because the multiple 3-state drivers can be connected to different power rails and different resistor combinations can be used.

The number of bits, N, can have two values for adcBIST. One value, N*out*, corresponds to the number of bits of resolution available from the ADC, for example, 12 bits. The second value, N*parallel*, corresponds to the number of bits output by the ADC each sampling clock cycle, and this can be less than or equal to N*out*, for example, 1 bit for the serial bit stream from a delta-sigma modulator (before decimation). Multi-bit delta-sigma modulators can be tested by adjusting the value of N*parallel*.

The number of master clock cycles of delay through the ADC must be provided to the BIST circuit generation software, and is typically between 1 and 32.

Primary Benefits

The following are the primary benefits of using BIST for ADCs:

- Automated generation of test patterns suitable for any tester
- Automated generation, insertion, and verification of the test circuitry
- Testing of ADCs using low-cost, digital ATE
- Faster, more thorough test than a digital ATE applying 2^N DC levels
- Testing at sample rates faster than the ATE data rate
- Testing at input frequencies higher than the ATE capability
- Reuse of the ADC test at the IC, board, and system levels

Measurements for ADCs

Each of the tests is individually enabled via the test access port, which is typically the 1149.1 TAP.

Best-Fit Polynomial

The Best-Fit Polynomial test measures offset, gain, and second and third harmonic distortion by applying the equivalent of an analog ramp to the input of the ADC while accumulating the ADC's digital output, and then calculating the best-fit third-order polynomial and the distortion that would be imparted to a rail-to-rail sinewave. Offset, gain, and harmonic distortion, as reported by adcBIST, are defined next.

The BIST circuit applies a modified analog ramp to the N-bit ADC, and finds the best-fit third-order polynomial. It also checks that the LSBs toggle.

Assume n samples are required, and that the ADC is clocked at its normal sample rate, f_S. Assume that the master clock frequency f_{CLK} is equal to f_S.

A modified ramp voltage is generated across a capacitor, C, driven by a D-type flip-flop (clocked by f_{CLK}), in series with a digital buffer and a resistor, R, as follows:

1. Output a logic 0 value for a time period S equal to n sampling clock periods.

2. Output logic 0 and 1 values, with a 25 percent duty cycle, for time period S.

3. Repeat step 2, each time changing the duty cycle: 50 percent, 75 percent.

4. Output logic 1 values for time period S.

5. Repeat step 2, each time changing the duty cycle: 75 percent, 50 percent, 25 percent.

6. Repeat steps 1 to 5, in sequence, as many times as necessary to allow AC-coupling to settle. (The default is zero.)

The resulting waveform is shown in Figure 6-7.

Figure 6-7 Typical ADC Input Voltage Generated by adcBIST

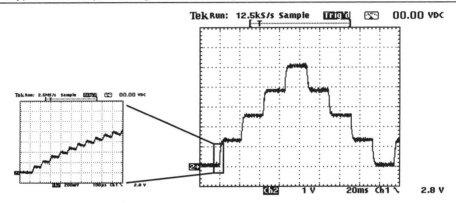

The output of the ADC is sampled and accumulated during the first $n/8$ sampling clock cycles of each step above, which is when the voltage across the capacitor is changing most rapidly. Four transitions occur going up (0 percent to 25 percent, then 25 percent to 50 percent and so on), and four transitions occur going down (100 percent to 75 percent, then 75 percent to 50 percent, and so on). At the end of each up transition, the accumulated value is stored in sequence in each of four registers, which can be labeled Sum_0, Sum_1, Sum_2, and Sum_3. For each of the down transitions, the accumulated value is added to the sum in the appropriate register. At the end of eight transitions (four up and four down), there are four sums whose value is almost independent of the RC value, and dependent on the offset, gain, and distortion in the ADC. The summing is slightly different depending on whether the ADC output coding is unsigned binary, sign plus magnitude, or two's complement, but the result is always in unsigned binary.

The relationship between the four sums and the four parameters of interest is as follows (assuming gain is approximately unity):

Offset $=$ $(Sum_3 + Sum_2 + Sum_1 + Sum_0) / n$

Gain $=$ $4 (Sum_3 + Sum_2 - Sum_1 - Sum_0) / n^2$

2nd harmonic $=$ $4 (Sum_3 - Sum_2 - Sum_1 + Sum_0) / n^2$

3rd harmonic $=$ $8 (Sum_3 - 3 Sum_2 + 3 Sum_1 - Sum_0) / 3n^2$

These equations are formally proven in Reference [7]. The principle is also described in U.S. Patent 5659312.

The binary-encoded digital solutions to the preceding equations (more specifically, the sums within brackets) are calculated by the BIST circuitry on-chip, and stored in four registers. To reduce the register size for the gain result, the gain error is calculated relative to unity gain.

An N-bit ADC with an LSB stuck at 0 might still be perfectly linear, and hence a check is needed to guard for this (in addition to the DNL test). While performing the accumulations, the BIST simultaneously checks that all of the LSBs toggle. If the LSB is stuck at 1 or 0, it would cause only a half LSB increase in the offset (and likely remain within test limits), and likely cause the DNL test to fail. By specifically testing for and reporting stuck bits, diagnosis is simplified. The default number of LSBs tested is equal to 3, or $N-5$, whichever is more. The number of LSBs tested can be set at compile time, and can be zero (to save approximately 50 logic gates).

Next, the binary-encoded, upper limits (scaled by a factor of n or $4/n^2$, and so on) for all four parameters are serially shifted into the IC, via the TAP. Next, the contents of the registers are shifted out, each value immediately followed by a pass/fail bit, followed by a code which indicates if any bits failed to toggle. This sequence is repeated for the lower limits. The tester needs only to check each pass/fail bit to determine whether the ADC passes the test. The measured value can be stored in the tester for later statistical analysis if desired. The shifted-in limits can be any value if the pass/fail bit is to be ignored, for example, during characterization.

When testing delta-sigma ADCs, the single-bit or multi-bit stream output of the modulator can be accessed directly. This is especially useful if the decimation is performed on another IC.

DNL

Differential non-linearity is the difference between the ideal voltage step size in the time-domain transfer function of the ADC, and the actual voltage step size, and is expressed in terms of LSB. BIST measures the maximum differential non-linearity (DNL) at user-selected MSB transitions (which could be all transitions). It does this by measuring the voltage difference between the analog switching points of two consecutive digital output values at each transition.

adcBIST applies a mostly cyclic waveform to the ADC, via the resistor and capacitor, and continuously adjusts the duty cycle to maintain the average value at the switching point of the ADC, for a particular output code. This is repeated for the next higher output code, and the difference in duty cycle is calculated. DNL tests are not needed for a delta-sigma ADC.

To simplify, this explanation assumes a 6-bit ADC. To save test time, DNL is measured only for transitions of the two most significant bits (MSBs). Therefore, 010000 would be shifted into the BIST circuit. The measurement requires the default +/-0.125 LSB resolution, that is, 9 bits.

The BIST circuit generates a digital waveform across the capacitor, C, driven by a D-type (clocked by f_{CLK}), in series with a digital buffer and the resistor, R, as follows:

1. The duty cycle of the waveform is adjusted so that the output of the ADC has an average value of 010000.

2. The duty cycle (that is, count the number of logic 1 values) is measured for a period equal to $2^9 = 512$ clock cycles; this number is A_1.

3. The duty cycle of the waveform is adjusted so that the output of the ADC has an average value of 001111.

4. The duty cycle, A_0, is measured for a period again equal to 512 clock cycles.

5. DNL_0 is equal to $(A_1 - A_0)/512$.

6. DNL_{MAX} is set equal to DNL_0.

7. Steps 1 through 5 are repeated for 100000 & 011111 (DNL_1); if $DNL_1 > DNL_{MAX}$ then DNL_{MAX} is set equal to DNL_1.

8. Steps 1 through 5 are repeated for 110000 & 101111 (DNL_2); if $DNL_2 > DNL_{MAX}$ then DNL_{MAX} is set equal to DNL_2.

The final value of DNL_{MAX} is the maximum DNL recorded at the selected MSB code transitions (which are most likely to be the worst).

Next, the binary-encoded upper limit for DNL is serially shifted into the IC via the TAP. Next, the value of DNL_{MAX} and the code at which it occurs are shifted out, immediately followed by a pass/fail bit.

BIST can measure DNL at any number of transitions. To measure DNL at transitions of the two most significant bits, as just described, 010000 is shifted in at test time. To measure DNL at the transitions of the three most significant bits, 001000 is shifted in. To measure DNL at every bit transition, 000001 is shifted in.

DNL cannot be tested if the ADC is AC-coupled to the BIST circuit.

Idle Channel Noise

Typically, idle channel noise is the RMS value of the digital output values when a zero-noise, mid-rail DC voltage is applied to the ADC. The BIST measures idle channel noise at mid-range by applying a cyclic AC waveform to the ADC and monitoring the ADC output whenever its input is at a specific DC value. In effect, a histogram is calculated and the user can select where to define the limits, for example, peak-to-peak or +/-1 sigma (twice the RMS value).

A cyclic waveform is applied to the ADC, and the noise is measured in the digital output at the same time point within each cycle, at which time the input is a constant mid-range DC voltage. This is a wideband measurement, and is generally not suitable for a delta-sigma bit stream; for delta-sigma converters, the decimated, band-limited output must be sampled for noise measurement.

The following examples assume a 6-bit ADC.

adcBIST drives a digital waveform into the resistor to produce a repetitive analog waveform at the input to the ADC. Relative to a mid-range voltage, a cumulative distribution function (CDF) of the output digital codes is produced, similar to the one shown in Figure 6-5 on page 145 for pllBIST.

For example, for the output code 100000, the number of times that codes less than that value are output might be 127 times out of 256, corresponding to 50 percent. For the output code 100001, the number might be 191, corresponding to 75 percent.

The resolution of this test is the voltage change in one master clock cycle. When $n = 2^N$, and $f_{CLK} = f_S$, the resolution is approximately 4 LSB. Table 6-2 shows the resolution for other combinations of n and f_{CLK}.

Table 6-2 Idle Channel Noise Measurement Resolution, in LSB

f_{CLK}	$n = 2^N$	$n = 2*2^N$	$n = 4*2^N$	$n = 8*2^N$
f_S	4 LSB	2 LSB	1 LSB	0.5 LSB
$2 f_S$	2	1	0.5	0.25
$4 f_S$	1	0.5	0.25	0.125
$8 f_S$	0.5	0.25	0.125	0.0625

The difference between the codes corresponding to 1 percent and 99 percent is approximately equal to the peak-to-peak noise, in units of LSB. The difference between the codes corresponding to 16 percent and 84 percent is approximately equal to twice the RMS noise.

Immediately before running the test, the specific percentage limits of interest are shifted into the IC, and the corresponding difference is calculated. After the test is completed, the binary-encoded upper limit for noise is serially shifted into the IC via the TAP. Next, the value of noise (as defined by the percentage limits) is shifted out, immediately followed by a pass/fail bit.

Values of R and C

The value of RC is calculated by BIST circuit generation software. Its value is $n/8f_S$, where n is the number of samples averaged for the N-bit ADC, and is usually equal to 2^{N+2}, and f_S is the sampling frequency of the ADC. Because the value of RC should be within 20 percent of this calculated value, the minimum value of R is constrained by the variation in the output impedance of the digital driver, and the minimum value of C is constrained by the unknown parasitic capacitances. Generally, R will be higher than 1 kΩ, and C will be larger than 100 pF. Table 6-3 on page 163 shows representative values. This table also compares the stimulus slew rate to that of a conventional sinewave (typically one-eighth of the sample clock frequency) used for Fast Fourier Transform (FFT) analysis. n can be smaller than 2^{N+2} to obtain a faster stimulus, or larger to reduce variance (improves repeatability).

Table 6-3 RC and Slew Rate Compared to Conventional Sinewave

f_S	n	R	C	max. slew for Vrange=1 $=n/8Rf_S$	equivalent sine wave $=Vrange/RC$	conventional sine wave $=1/\pi RC$	$= f_S/8$
(MHz)	samples	(kΩ)	(pF)	(V/μs)	(MHz)	(MHz)	
50	64	1	160	6.3	2.0	6.3	
100	64	1	80	12.5	4.0	12.5	
250	64	1	32	31.3	9.9	31.3	
500	64	0.5	32	62.5	19.9	62.5	
50	256	5	128	1.6	0.5	6.3	
100	256	5	64	3.1	1.0	12.5	
200	256	5	32	6.3	2.0	25.0	
500	256	2	32	15.6	5.0	62.5	
20	1024	10	640	0.2	0.05	2.5	
50	1024	10	256	0.4	0.12	6.3	
100	1024	10	128	0.8	0.25	12.5	
200	1024	5	128	1.56	0.50	25.0	
10	4096	20	2560	0.02	0.01	1.3	
20	4096	20	1280	0.04	0.01	2.5	
50	4096	20	512	0.10	0.03	6.3	
100	4096	10	512	0.20	0.06	12.5	

The value of *R* must include the output impedance of the digital driver. Pin drivers for a typical IC have user-selected output impedances between 25 and 100 ohms, varying by +/-40 percent over temperature and process (variation in V_{DD} does not need to be accounted for if the ADC and digital buffer have equal sensitivities). As well, there will be a difference between the n-channel transistor drive and the p-channel transistor drive. To make all these variations negligible, the value of the resistor that is in series with this output impedance must be relatively large.

The value of *C* must include any parasitic capacitances. An IC pin and associated printed board wiring can easily contribute 2 to 10 pF, which must be accounted for. Fortunately, its value will be fairly constant over all conditions.

If the value of *RC* deviates more than 20 percent from the value calculated by software, then the measurement results will "gracefully" become less accurate. When RC is too small, some regions of the ADC's voltage range (at 1/8, 3/8, 5/8, and 7/8) are counted twice, thus non-linearity at those voltages will be exaggerated when calculating harmonic distortion. When RC is too large, nonlinearities at these same points will not be counted at all, thus diminishing the contribution to harmonic distortion. Within 50 percent of the correct RC value, when N=12, adcBIST will report the correct offset and gain; "zero" distortion will still be reported as zero distortion; nonzero distortion measurement will be inaccurate.

Looking at a sample 8-bit ADC sampling at 50 MHz, the value of *RC*, from Table 6-3, is 10 kΩ and 256 pF (RC=2.56 μs), assuming 1024 samples are used to calculate the distortion. If the parasitic capacitance is known to be between 4 and 11 pF, then this amounts to 7+/-3 pF. An off-the-shelf 100 pF +/-10 percent capacitor can be used to achieve 107 pF +/-13 percent.

The resistance needs to be 640 nanoseconds divided by 107 pF, which equals 5.98 kΩ. A +/-7 percent variation can be tolerated, which is equal to +/-420 ohms. Using off-the-shelf 1 percent resistors driven by a 100 +/-50 ohm digital driver, a +/-2 percent can be achieved, which easily meets this requirement. If an on-chip resistor is used, then +/-7 percent is difficult to achieve, so a 1-percent capacitor is needed.

The simplest implementation of BIST occurs when the output voltage range of the digital buffer is the same as the input voltage range of the ADC. This can often be arranged by use of a level shifter, so that the Vout signal from the BIST circuit is level-shifted by a (3-state) digital buffer whose voltage rails are the same as the ADC's input.

When the input voltage range of the ADC is less than the voltage swing of the digital output, then a resistive divider circuit must be used. There are three possibilities:

Case A—ADC voltage range is from V_{SS} (0 V) to some value less than V_{DD} (for example, 5 V).

Case B—ADC voltage range is from V_{DD} to some value greater than V_{SS}.

Case C—ADC voltage range is from some value greater than V_{SS} (0 V) to some value less than V_{DD}.

Case A: VSS Less Than VDD

For Case A, ADC voltage range is from V_{SS} (0 V) to a value less than V_{DD} (for example, 5 V). Figure 6-8 shows the resistor divider that must be used for Case A.

The ratio of R_1 to R_2 must be more accurate than the required gain accuracy of the ADC, however the value of the equivalent output resistance times the capacitance need only achieve +/-20 percent accuracy. The value of R_1 must include the output resistance of the digital driver. The equivalent resistance is R_1 in parallel with R_2, which equals $R_1R_2 / (R_1+R_2)$.

Figure 6-8 Circuit for Scaling the Maximum Output Voltage from BIST Digital Circuitry

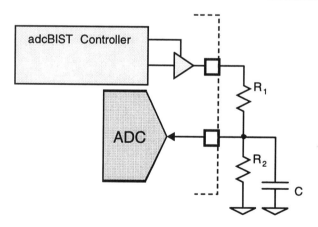

Case B: VDD Greater Than VSS

For Case B, ADC voltage range is from V_{DD} to a value greater than V_{SS}. The circuit of Figure 6-8 is used, but R_2 is connected to V_{DD} instead of V_{SS}.

Case C: Greater Than VSS to Less Than VDD

For Case C, ADC voltage range is from a value greater than V_{SS} (0 V) to some value less than V_{DD}. The circuit of Figure 6-9 is used to achieve an arbitrary voltage range, with an arbitrary maximum output level and arbitrary minimum output level.

Figure 6-9 Circuit for Scaling the Min/Max Output Voltages from BIST Digital Circuitry

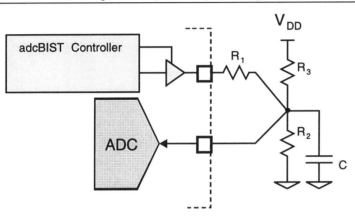

Again, the ratio of R_1 to R_2 to R_3 must be more accurate than the required gain accuracy of the ADC. The equivalent resistance is R1 in parallel with R_2 in parallel with R_3, which equals $R_1R_2R_3/(R_1+R_2+R_3)$.

The stimulus for an ADC with a differential input is generated using the circuit of Figure 6-10. Voltage scaling, which is typically needed for this application, is provided using R_2.

Figure 6-10 Circuit for Scaling the Min/Max Output Voltages from BIST Digital Circuitry

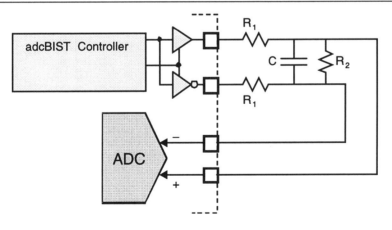

Values for *n* (Number of Samples)

The variation (or more mathematically, the variance) in the measurement result is due to random noise and quantization noise, and is inversely proportional to *n* (or a power of *n*). To improve test repeatability, which allows smaller guardbands or higher yield, the variance needs to be minimized. However, taking more samples increases test time. As seen in Table 6-4 on page 167, taking more samples also decreases the slew rate of the stimulus ramp because a larger RC value will be used.

For an *N*-bit ADC, the default choice for *n* is $n=2^{N+2}$. This is a compromise between maximal noise tolerance and minimum test time. For ADCs with sample rates above 20 MS/s, the test time for 1024 samples is only 0.4 ms, so increasing this test time by 16 times is relatively insignificant, but it reduces the RMS variation in the result by a factor of four.

In some applications, other analog circuitry will contribute offset and gain errors, or non-linearity, so using more and more output samples from the ADC will have diminishing return in overall system quality. The linearity of the RC also imposes a fundamental limit on the linearity of adcBIST. More than 16384 samples will probably provide insignificant improvement in test quality, although the software for creating BIST will allow almost any number of samples that is a power of 2 (it requires more logic gates).

A larger value of *n* also allows finer resolution in the noise test. Choosing the default value of $n=2^{N+2}$ gives approximately 1 LSB resolution when the master clock frequency is equal to the sampling clock frequency. Increasing or decreasing *n* changes the resolution proportionately. If the master clock frequency is higher than the sampling clock, then the noise resolution will be proportionately finer too.

Because the ADC is operated at full speed, many speed-related effects are covered by the test. However, the analog stimulus is relatively slow compared to the maximum input slew rate that the ADC must accommodate. Choosing a smaller value of *n* will allow a faster stimulus slew rate, though reducing *n* to 64 still results in a slew rate significantly slower than the ADC's maximum rate (the slew rate used in the DNL test can be much faster because a much smaller resistance can be connected during the test).

Test Times

This section summarizes the test times for the three tests.

Best-Fit Polynomial

This test requires 8*n* sampling clock cycles to perform a single ramp up/down sequence, so if $n=2^{N+2}$ is chosen, which is the default, 2^{N+5} sampling clock cycles are required.

For example, for a 10-bit, 20-MS/s ADC for which 4096 samples are accumulated, the test time would be 1.7 ms (plus the time it takes to initiate the test and read the results via the TAP).

If the ADC is AC-coupled, then the up/down ramp is repeated continuously until the AC-coupling has settled (a constant interval, programmed by the user), and the test time is increased by this settling time.

DNL

This test requires 2^{N+4} clock cycles per DNL measurement (two switching points, 1/8 LSB resolution), so if the 7 most significant transitions are measured (that is, for the three most significant bits), the test time will be $7*2^{N+4}$.

For example, for the 10-bit, 20-MS/s ADC, for which DNL is tested at transitions of the 3 MSBs, the test time would be 5.7 ms. To measure DNL for all code transitions, the test time would be 840 ms. Measuring DNL for all code transitions may be too long for a production test when the number of bits is greater than 8, but is very useful during characterization and for a few devices per wafer or lot.

Idle Channel Noise

This test requires E*32K clock cycles, where E is the maximum expected peak-to-peak noise in LSB (16 is default). If only RMS noise will be measured, then this test time could be reduced by a factor of 4.

For example, the 10-bit, 20-MS/s ADC, with less than 16 LSB peak-to-peak noise, will require 26.2 ms to test.

Total Test Times

In summary, for the sample 10-bit, 20-MS/s ADC:

Best-Fit Polynomial	1.6	(offset, gain, distortion)
DNL	5.7	(DNL_{MAX})
Idle Channel Noise	26.2	(RMS noise)
Total	**33.5 ms**	

Test times for a variety of resolutions and sampling rates are shown in Table 6-4.

Table 6-4 Test Times for Various ADC Resolutions and Sampling Rates

N	f_S	n 2^{N+2}	Poly $8n/f_S$	DNL $7*2^{N+4}$	Noise $16*32K$	Total	DNL, all transitions
bit	(MHz)	samples	(ms)	(ms)	(ms)	(ms)	(ms)
6	50	256	0.04	0.143	10.5	11	1.33
6	100	256	0.02	0.072	5.2	5	0.66
6	200	256	0.01	0.036	2.6	3	0.32
6	500	256	0.00	0.014	1.0	1	0.14
8	20	1024	0.41	1.434	26.2	28	52
8	50	1024	0.16	0.573	10.5	11	20
8	100	1024	0.08	0.287	5.2	6	10
8	200	1024	0.04	0.143	2.6	3	6
10	10	4096	3.28	11.469	52.4	67	1678
10	20	4096	1.64	5.734	26.2	34	838
10	50	4096	0.66	2.294	10.5	13	336
10	100	4096	0.33	1.147	5.2	7	168
12	5	16384	26.2	91.750	104.9	223	53,688
12	10	16384	13.1	45.875	52.4	111	26,844
12	20	16384	6.55	22.938	26.2	56	13,422
12	50	16384	2.62	9.175	10.5	22	5,368
		resolution =		0.125	1 LSB		0.125 LSB

Typically, 4096 or 8192 samples will provide sufficient measurement accuracy for ADCs that have more than 12 bits of resolution. This will mean that fewer than 2^N samples are used for an N-bit ADC, so not all output codes will be tested. This is common practice—for example, the number of samples used for a Fast Fourier Transform (FFT) is typically 2048 or less because of the computational complexity and resulting impact on test time. For both adcBIST and the FFT, the fault coverage of the test remains high because the input voltage is uncorrelated to the quantization intervals.

Assumptions for Testing ADCs

To economically achieve the required accuracy and resolution, several assumptions are made about the N-bit ADC to be tested, and the environment in which it is located. If these assumptions are not already valid for an IC design, the following guidelines are provided to help make the assumptions valid.

The assumptions are as follows:

- The minimum/maximum voltage decoded by the ADC is the same as the output logic 0/1 voltage. If these voltages are not the same, a resistor network can be used when the ADC voltage range is smaller.

- The impedance variation of the BIST circuit's digital output, for logic 0 and 1 voltages, is negligible relative to the value of R. If this is not, the resistor's value can be made suitably high.

- The input impedance of the ADC is much higher than R. If this impedance is not higher than R, an analog buffer can be inserted between the ADC and the capacitor.

- A $2N$-bit accumulator can be synthesized to be clocked at the sample rate of the ADC.

- The linearity of R and C, with respect to voltage, is better than the ADC to test; that is, the value of RC does not vary over the voltage range of interest by more than $1/2^N$ times the nominal value of RC. If this linearity is not better than the ADC under test, an off-chip, discrete resistor and capacitor can be used.

- The value of RC prescribed for the BIST can be achieved, within appropriate accuracy and on-chip area constraints. If this value cannot be achieved, an off-chip discrete resistor and capacitor can be used, and/or a suitably slower stimulus can be generated by the BIST circuit.

- When an off-chip RC or C is used, as is often the case, off-chip access to the analog input of the ADC should not corrupt the ADC's performance.

- Means are available for connecting the RC output to the ADC's input, and disconnecting the normal stimulus (for example, an analog multiplexer or tri-state buffers).

- A single ADC is tested. If multiple ADCs are tested, a register and multiplexers external to the BIST controller can be added to enable testing of more than one ADC.

- An at-speed sampling clock is available.

References

[1] F.M. Gardner, *Phaselock Techniques*. 2nd edition, John Wiley, New York, 1979.

[2] F.M. Gardner, "Charge-Pump Phase-Lock Loops," In *IEEE Trans. on Communications*, vol. 28, no. 11, pp.1849-1857, Nov. 1980.

[3] P.V. Brennan, *Phase-Locked Loops: Principles and Practice*. MacMillan Press Ltd., Great Britain, 1996.

[4] S. Sunter and A. Roy, "BIST for Phase-Locked Loops in Digital Applications." in *Proceedings of the IEEE International Test Conference (ITC)*, September 1999.

[5] M. Mahoney, *DSP-Based Testing of Analog and Mixed-Signal Circuits*. IEEE Computer Society Press, 1987.

[6] S.R. Norsworthy, R.Schreier, G.C.Temes, *Delta-Sigma Data Converters: Theory, Design, and Simulation*. New York, IEEE Press, 1996.

[7] S. Sunter and N. Nagi, "A Simplified Polynomial-Fitting Algorithm for DAC and ADC BIST," in *Proceedings of the IEEE International Test Conference (ITC)*, November 1997.

Chapter 7
System Test and Diagnosis

This chapter describes three technologies used to test and debug boards and systems.

The first section describes an interconnect test capability that enhances the IEEE 1149.1 TAP and boundary-scan cells standard to provide built-in timing control so that complete at-speed testing of board-level interconnect is achieved. The approach is completely new and fully automated.

The second section describes an automated solution for testing off-the-shelf memory components. On average, memory accounts for about 30 percent of the components used within a system. A board and system-level test strategy cannot be complete without a test strategy for memory components or modules.

The third section describes the *fault insertion* capability. This technology is used to verify functional error detection, recovery, and diagnostic mechanisms built into the hardware and software of the system being tested. Fault insertion is built on top of the existing boundary-scan infrastructure and benefits from the same level of automation provided by the boundary-scan insertion tools.

At-Speed Interconnect Testing

One of the most significant challenges of the electronics revolution is the requirement to adequately test complex, high-performance chips and boards in order to maintain the near-perfect quality level expected by customers. To meet these expectations, as well as time-to-market and cost requirements, organizations must thoroughly plan test schemes for chips, boards, and systems. Organizations with foresight are planning unified, hierarchical test strategies that span chip, board, and system-level manufacturing. Two methods that can serve as the foundation for these test strategies are BIST and boundary-scan test.

Boundary-Scan Test Strategy

BIST, or embedded test, is typically controlled by a TAP, which is a five- pin test interface whose operation and use is defined in the IEEE 1149.1 JTAG standard. The TAP's five pins comprise the following:

- Serial data input (TDI) for shifting in test and control data
- Serial data output (TDO) for shifting out test responses
- Test clock input (TCK)
- Test mode input (TMS)
- Optional reset input (TRST)

These five pins can be used with a simple tester to send a few short instructions to the TAP. The instructions are channeled inside a chip to the appropriate BIST controller to initialize the controller, start the BIST, and capture the pass/fail result.

Although designed to support the access of chip self-test mechanisms, the primary purpose of the IEEE 1149.1 JTAG standard is to provide an effective alternative to probe-based, board-level interconnect testing methodologies. To accomplish this, the standard defines, in addition to the TAP, a boundary-scan architecture to be implemented in all chips involved in interconnect tests. Boundary scan is enabled by adding a shift-register stage (referred to as a boundary-scan cell) adjacent to each chip pin so that signals at the chip's boundaries can be controlled and observed using scan-testing principles.

These boundary-scan cells are then interconnected to form a shift-register chain around the periphery of the chip. This boundary-scan chain is accessed serially through the TDI and TDO pins of the TAP. At the board level, the boundary-scan chains of individual chips are serially connected to form a single board-level chain.

To test the interconnect between chips, test data can be shifted into all the boundary-scan cells associated with chip-output pins and loaded in parallel through the chip interconnections into boundary-scan cells associated with chip-input pins. The captured data can then be shifted out and compared to the expected result. When the correct data is not captured at a pin of the receiving chip, the interconnect segment between the two devices is suspect.

Unfortunately, the IEEE 1149.1 test approach cannot apply and capture data across interconnects at application speed for boards manufactured today. This disadvantage causes delay defects to be missed. A delay defect is a flaw on a board that prolongs the travel time of a signal along an interconnect line. Examples of delay defects include board imperfections such as cold solder joints, thin or narrow regions of interconnect, or thin insulation between levels of interconnect. These defects cause the interconnect resistance or capacitance to be higher than designed, increasing the time required for a signal to move from the *sending* chip to the *receiving* chip.

For IEEE 1149.1-based testing, a minimum of 2.5 TCK cycles are required between the time test during which data can be launched from one chip and captured at another. Because TCK operation is typically limited to operating frequencies of 2 to 20 MHz, the effective interconnect test frequency available from 1149.1 is from 800 KHz to 8 MHz, which is much lower than typical board application speeds.

Existing At-Speed Interconnect Test Strategies

Many companies today perform interconnect testing by using in-circuit test (ICT) techniques. Testers that provide this capability typically contain hundreds to thousands of physical probes (*bed of nails*) that are used to contact chip pins and board testpoints to perform the interconnect tests.

For today's designs, however, the use of ICT is often limited because of the total number of probes required and the physical limitations of accessing chip pins due to overall board densities and packaging techniques such as ball grid arrays (BGAs). In addition, testing for interconnect delay faults is not practical because of the inability of the tester hardware to apply at-speed signals to hundreds or thousands of probes.

The method most commonly used to achieve at-speed interconnect testing is functional testing at the system level. However, when an error is detected using this approach, it can take days or weeks to diagnose the precise location of the problem. Obviously, a better method of testing interconnect at application speeds is required to ensure that highly complex systems can be taken to market at the appropriate time and that production deliveries and costs can be kept under control.

New Solution

LogicVision's patented jtag-XLi technology enables designers to apply and capture data across board interconnects during one system clock cycle. This at-speed interconnect technology builds upon the standard IEEE 1149.1 JTAG board test approach. It uses a combination of new at-speed boundary-scan cells as well as an ASI test controller to drive necessary timing signals to these new cells. The at-speed boundary-scan cells consist of standard 1149.1 cells augmented with specialized logic and a system-clocked flip-flop. The cells are still scanned directly from the TAP using the TCK test clock. Instead of being controlled by the standard update and capture signals generated by the TAP controller, the at-speed boundary-scan cells are controlled by at-speed versions of these signals generated by the embedded ASI test controller. A board-level synchronization signal is routed between the ASI test controllers residing on the chips involved in the at-speed interconnect test. This signal ensures that the operation of ASI controllers on a board are all aligned to the same system clock cycle. The synchronization signal can be shared with a functional signal to avoid adding any more I/O to the device. Figure 7-1 illustrates the chip-level architecture required to implement the at-speed interconnect technique.

Figure 7-1 Chip-Level Architecture for At-Speed Interconnect Testing

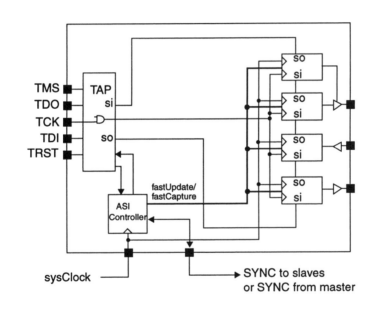

Enhanced IEEE 1149.1 for At-Speed Testing

One of the key advantages of the at-speed interconnect approach is its complete transparency to IEEE 1149.1-based testing. The board design can include a mixture of chips that incorporate the at-speed interconnect technology with other chips that are compliant with the IEEE 1149.1 standard. Any existing boundary-scan test hardware, as well as test and diagnostic software tools, can be used with both chip types without modification. In this scenario, only the interconnect between chips that incorporate the at-speed technology is tested at speed; the remaining interconnect is tested at standard TCK rates.

Figure 7-2 illustrates the difference between IEEE 1149.1 standard testing and the at-speed testing approach. As mentioned earlier, the problem with IEEE 1149.1 testing is that a minimum of 2.5 TCK cycles are required between the time test data is launched from one chip and received at another. The at-speed test approach reduces this time to one system clock cycle, providing complete at-speed testing of the interconnect.

The test is also representative of the normal mode of operation because functional clocks and flip-flops determine the timing. Therefore, subtle timing problems such as clock skew between chips can be detected.

Figure 7-2 IEEE 1149.1 Versus At-Speed Interconnect Testing

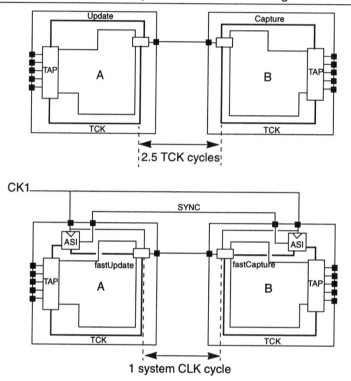

At-Speed Boundary-Scan Cells

The at-speed boundary-scan cells consist of standard 1149.1 cells augmented with specialized logic. They also use an existing system-clocked flip-flop associated with an input or output pad or include this flip-flop in the cell, if necessary. Although the cells are scanned directly from the TAP using the TCK test clock, they are designed to launch and capture data across the interconnect using the system clock.

Examples of output and input cells are illustrated in Figure 7-3 and Figure 7-4 on page 176. The functional input and output signals are registered using the system clock. This is very frequent in high-speed systems to minimize the spread of timing characteristics caused by process variations of the integrated circuits. The modified boundary-scan cell still supports the standard instructions of 1149.1. In particular, during a test of the board interconnect, the value shifted in the output boundary-scan cell is applied to the corresponding output pad upon entering the *Update_DR* state of the TAP.

Figure 7-3 Boundary-Scan Output Cell with At-Speed Test Support

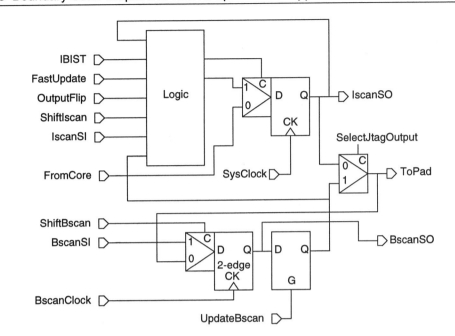

In the *at-speed* mode, the value is applied instead to the input of the functional register and released at an appropriate time using the *FastUpdate* signal. This signal is generated by the ASI controller as well as the *IBIST* and *OutputFlip* signals. *IBIST* is a static signal that indicates that the circuit operates in the at-speed mode. *OutputFlip* is an optional signal that is used to test the hold time of the receiving flip-flops in other chips.

On the input side, the pad value is captured at the time indicated by the *FastCapture* signal, also generated by the controller, in the at-speed mode. The input boundary scan cell also supports the standard 1149.1 operations.

Figure 7-4 Boundary-Scan Input Cell with At-Speed Test Support

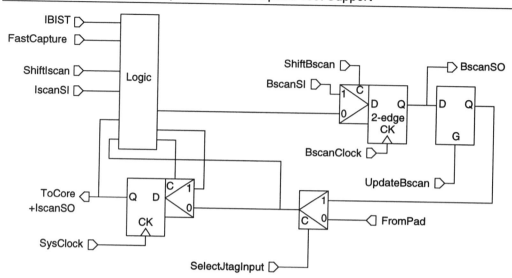

The *ShiftIscan*, *IscanSI*, and *IscanSO* signals are used for internal scan purposes and are already present. In a muxed-scan implementation such as the one in Figure 7-3 on page 175, there is no performance impact when the at-speed test is implemented because logic is added on the scan path instead of the functional path. Only a small amount of circuitry is required to implement the new cells. Variations of these cells can be generated for bidirectional pads and non-registered inputs and outputs.

ASI Controller

Each ASI controller on a board is responsible for providing an at-speed update signal and a capture signal to all at-speed boundary-scan cells within a common clock domain. Each ASI controller also sends or receives a board-level sync signal so that all at-speed controllers within a common clock domain are aligned to the same system-clock cycles. The TAP is designed to control the ASI controller and to provide access to the at-speed boundary-scan cells and any other on-chip test or embedded ATE capabilities.

At-Speed Interconnect Timing

Figure 7-5 on page 177 shows a timing waveform for at-speed interconnect. A key point with this timing example is that the at-speed update and capture sequence occurs between the slow update and capture events. This sequence makes the at-speed technique fully transparent to 1149.1 operation.

Another key point is that the signal driven on the at-speed output pin is inverted for one clock cycle immediately after the at-speed capture event. This sequence tests for hold-time violations on the receiving flop.

Figure 7-5 At-Speed Interconnect Timing

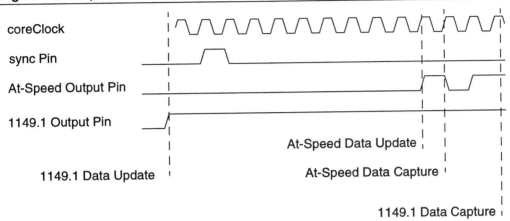

Design Flow

A Generate tool processes design-related information about the I/Os of the target chip, target inter-connects, and other user-specified options. (An example of a user-specified option is the frequency of the system clocks used for the test.) Based on this information, the tool outputs synthesizable RTL design objects and synthesis scripts for each ASI controller. There is one controller for each functional clock domain.

Controllers are added at the top level of the chip at the same time as the TAP and boundary-scan cells, during top-level flow described in Chapter 4, "Embedded Test Design Flow."

A Verify tool creates testbenches to verify the functionality of the ASI controllers at the design phase. Test and diagnostic patterns are generated in SVF to provide links to system test.

At-Speed Memory Testing

This section provides background information about off-the-shelf memories and strategies for testing these memories. A new solution that allows at-speed testing of off-the-shelf memories is presented along with the tools to support the implementation. The solution also enables designers to test and diagnose the interconnect between the custom integrated circuit that contains the embedded test controller and the memory components or modules.

A significant challenge at the board and system levels is testing external memories. Commercial external memories, widely used in complex electronic products, typically lack built-in capabilities designed to make them easily testable at board and system levels. Furthermore, several attributes contribute to the difficulty of testing external memories: *size, test complexity and time, packaging, type,* and *organization.* The problems associated with each attribute are described below.

- *Size*—The amount of memory within a board or a product ranges from a few kilobytes to several hundred gigabytes.

- *Test complexity and time*—The density and resulting complex failure modes make memory test difficult. Theoretically, the time required to fully test a memory grows exponentially with size. Although efficient, linear algorithms that target real failure modes are available, testing large memories can take unacceptable amounts of time, especially if testing is controlled by an external tester.

- *Packaging*—Because external memories are typically used in clusters in larger memory systems, they are prime candidates for high-density interconnect and packaging technologies such as dual-sided surface mount, MCMs, direct chip attach, and 3-D packaging. Memory modules such as SIMMs (Single-Inline Memory Module) and DIMMs (Dual-Inline Memory Module) attach to the board via connectors. This variety makes physical access difficult.

- *Type*—External memories come in a wide range of types: static, dynamic, different access mechanisms; for example, EDO (Extended Data Out) and synchronous burst, and single or multi-port. This variety makes test development difficult.

- *Organization*—Memory banks can have many different organizations based on the type of device used to construct the memory. The test and diagnostic method must comprehend this organization so that the appropriate level of diagnostics is provided.

A complete structural test of a board requires verifying that the correct devices are assembled properly and that no manufacturing defects were introduced by the assembly process. At the system level, the test must ensure all boards or modules are functional and other system-level interconnect (backplanes and cables) are defect-free. Ideally, all tests must be applied at the system's operating speed and environmental conditions to ensure that performance-related defects are detected and diagnosed.

The required degree of diagnostic accuracy and resolution varies. The highest degree is required during manufacturing test when a precise device, pin, or interconnect needs to be identified in order to facilitate repair. In the field, unambiguous identification to the field replaceable unit, typically a board, is the minimum requirement. A test methodology that provides the correct level of diagnostics is critical.

Finally, the return on investment (ROI) in off-line structural test can be substantially greater if the capability can support functional test, online test, and remote access, and can be reused for normal system functions such as reset, initialization, and programming.

Conventional Test Approaches

The two conventional test approaches—ICT equipment and IEEE 1149.1 JTAG—that are frequently used to test memory controllers and their associated memories on a board typically cannot guarantee that the board assembly is free of defects at application speeds.

Some of the more subtle board defects, such as cold solder joints, non-homogeneous strip-line width, or thickness, are characterized by abnormally high interconnect resistance and capacitance. These characteristics cause *interconnect delay*—the time that it takes for a signal sent from one chip to arrive at a receiving chip—to be longer than expected. Late arrival of signals at their destination on a board can cause failures during board or system-level functional test that are very difficult to diagnose. Structural tests of board interconnect at rated application speed allow interconnect delay defects to be located and easily diagnosed.

Test Approach: ICT Equipment

The ICT equipment approach uses an external tester to force test signals directly onto board strip-lines that are connected as follows:

- To the memory controller outputs.
- To capture the responses of each memory component by electrically contacting the strip-lines of its outputs.

ICT equipment that uses large numbers of metallic probes to contact board interconnect are commonly used to perform this test operation. Installed ICT equipment is often not designed to apply large numbers of signals at application speeds or to adequately control the timing between these signals.

Using ICTs to verify memory buses on complex boards often extends board layout design time and test development time. If application-speed testing is not done due to limitations of installed ICT equipment, subtle, difficult-to-diagnose problems caused by interconnect delay defects can arise during system functional test. At the system level, these types of problems can take hours, or even days, to diagnose and correct.

IEEE 1149.1 for At-Speed External Memory Testing

The IEEE 1149.1 test approach uses an external tester to scan test signals to the memory controller outputs using an IEEE 1149.1 TAP and boundary-scan register embedded in the ASIC that contains the controller. The output responses of each memory are captured by an IEEE 1149.1 TAP and boundary-scan register that is embedded in one or more components driven by the memory.

When the proper devices on a board contain an IEEE 1149.1 TAP and boundary-scan register, this approach can be used to perform a low-speed test of the interconnect between the memory controller and memory components. The maximum frequency of the test is limited by the frequency of the TAP's TCK and characteristics of the IEEE 1149.1 standard. It requires 2.5 TCK periods to perform a launch-and-capture operation to check the interconnect between the memory controller and the memory array.

A significant percentage of today's applications operate at a much higher frequency than that achievable with an IEEE1149.1 test scheme. Therefore, delay faults that require significant resources to diagnose at board and system-level functional testing steps typically go undetected.

Problem Summary and Solution

Both approaches described in the previous sections are burdened with other disadvantages and limitations.

- With significant advances in multi-layered board technology and rapid adoption of new high pin-count BGA packaging, board designers must include many testpoints to allow continued use of installed ICT equipment. This testpoint requirement complicates the board design because testpoints must be routed to the top-board level as part of the overall layout scheme.

- If testpoints on a board are moved because of a required change in the location or composition of the board's components, corresponding test program modifications are typically required.

- Test program development to verify proper operation of the memory controller and its memory array, while straightforward, is often tedious and time consuming. To verify the board connections between the controller and the memory components, the test program must check for stuck-at faults at all address, data, clock, and control pins of each memory component.

- Thorough retest of each memory component, to ensure that it is correctly functioning after assembly onto the board, requires application of a *March* test algorithm. (Refer to the "Memory Test Strategy" section on page 182.) If the installed ICT equipment does not contain memory test hardware, creation of this test routine requires significant programming resource. Retest of the memory components with an IEEE 1149.1 scheme is not practical. The number of tests required is very large, and each test vector must be serially shifted into the boundary-scan register of the memory controller device.

- For systems that require continuous *in-situ* test during use, the ICT approach must be replaced with a new system-level scheme. This scheme mandates a totally independent test development, typically requiring significant time and resource. If an IEEE 1149.1 approach is used during system manufacture, the *in-situ* test can be more easily implemented, but it cannot be a thorough test.

With the rapid advancement of technology in VDSM integrated circuits, a reliable test strategy is urgently needed to test clusters of memories on printed-circuit boards or in multi-chip modules. A complete, automated solution for testing external (off-chip) memory structures in hierarchical assemblies is available. LogicVision's memBIST-XT tool embeds design objects into a controller ASIC that test external memories. These design objects are created as RTL descriptions.

The memBIST-XT design objects include a test controller and an interface module.

- The controller is responsible for executing the test operation that includes the generation of test data and expected responses and the capture and compare of actual responses. The controller can be manipulated either from an 1149.1 TAP controller on the board or directly through pins located on the host ASIC.

- The interface module handles the interface between the memBIST-XT controller and memories on the board.

memBIST-XT Features

Appropriate design-related information must be provided to memBIST-XT about the target memory cluster. Based on this information, memBIST-XT generates the following:

- memBIST-XT controller design objects in synthesizable Verilog or VHDL RTL descriptions
- Memory interface logic design objects in synthesizable Verilog or VHDL RTL descriptions
- Verilog or VHDL testbenches to verify the functionality of the controller and interface structures in simulation
- Synthesis scripts

The test controller provides comprehensive testing of the external memories and of the interconnect between the controller ASIC and the memories. The controller is initialized and its test is started by simple instructions entered through the TDI of an IEEE 1149.1 test bus on a board. Designed for efficient, application-speed testing of the interconnect between a memory control bus and its memory components, a special interconnect test is automatically generated by the embedded test controller.

The controller can also apply *March* algorithms at application speeds to each of the memory components connected to the memory controller device. The controller can be configured to perform *go/no-go* testing. Additionally, a *stop-on-nth-error* test routine can be included that allows the capture of address and data failure information to support diagnostic work.

The controller is typically connected to an IEEE 1149.1 TAP so that the initiating the external memory self-test, obtaining the response and obtaining the diagnostics can be accomplished through this standard interface. This scheme permits testing of the memory to be integrated as part of the overall product test strategy. Access to the embedded tester can be preserved from board to system and can be reused in lab debug, manufacturing, field test, and diagnosis.

Memories can be fully tested at all phases of the product integration hierarchy and life cycle.

The external memory test solution described above provides the following benefits:

- The testing approach provides test speeds several orders of magnitude faster than traditional approaches that rely on the use of an external tester or embedded processor to test memory.
- Diagnostic features reduce debug time by quickly identifying the failing memory chips and/or interconnect. memBIST-XT allows the selection of *go/no-go* or comprehensive data capture for diagnostic support. The *stop-on-nth-error* feature allows the capture of sufficient failure data to allow complex analysis to highlight suspect segments of interconnect on a board or an MCM.
- Testpoints are not required to implement the solution, simplifying the work of the board designer.
- Test programming time is minimized. The test algorithms for interconnect and memory component testing are automatically generated by the test controller circuitry. The IEEE 1149.1 test routine required to run the controller can also be automatically generated.
- Interconnect delay faults are checked because all tests are at application speeds. This feature minimizes costly and time-consuming diagnosis at subsequent board or system-level functional testing operations.

- Memory components are thoroughly tested at application speeds after they are installed on the board. This procedure identifies defective memory devices that escaped detection during component test or were damaged during board assembly by electrostatic discharge.

- Improved design productivity results from automatic creation and hookup of the design objects for the controller and memory interface logic and of the verification testbench. An alternative approach of creating the instructions for a microprocessor to provide the test would likely require a larger development effort as well.

During test development, the controller can be configured to comprehend all features that are planned to be used. During actual board test, any of these features can be selected by providing the appropriate initialization data string to the board IEEE 1149.1 test bus. This selection can be done with a simple change to the test program of the external tester.

The options that can be exercised are as follows:

- Revise the definition of the memory cluster to comprehend multiple memory configurations. For example, the board is configured with different amounts of DRAM.

- Omit certain tests from the test routine. For example, when a low level of memory-component failures occur after board assembly, memory-component testing can be omitted and only the board-interconnect test can be performed, thereby significantly reducing test time.

- Select *stop-on-nth-error* test mode versus the *go/no-go* mode in order to collect failure data for diagnostics.

Memory Test Strategy

The controller implements several variations of a *March*-like algorithm with programmable read and write access sequences. This programmability allows the controller to interface with a wide variety of memory types (SRAMs, DRAMs, and SDRAMs).

The algorithms test for stuck-at-faults, shorts, and coupling faults for bit lines and data and address buses. The algorithms also test for cell-related faults such as stuck-at faults. For diagnosis, a *stop-on-nth-error* option allows scanning-out the failing address and data values through the controller and interface before resuming the test. It also supports the generation of a *result array*. A result array provides a map of failed interconnects or a map of failing memory chips within an address range.

The following steps provide a recommended test flow:

1. Use the *Interconnect* algorithm to check for faults in data, address, and control signals.

2. Use the *March* algorithm to check for pattern disturbs and to check for correct functional operation of each memory in the external array.

3. Diagnose memory failures by using the *result array* and *stop-on-nth-error* features.

Algorithms

The controller is configured based on one of the following six *March*-type test algorithms that can be selected.

- *March4N*—detects the following failure modes: stuck-at, address decoder, read/write logic, and parametric faults.

- *March4NJ*—March 4N with Johnson Counter background. This algorithm detects the same failure modes as the March4N algorithm. It also includes detection of bit-line coupling faults.

- *March 6N*—detects the following failure modes: stuck-at, transition, unlinked inversion coupling, address decoder, read/write logic, parametric, and write recovery faults.

- *March 6NJ*—March 6N with Johnson Counter background. This algorithm detects the same failure modes as the March6N algorithm. It also includes detection of bit-line coupling faults.

- *March 10N*—detects the following failure modes: stuck-at, transition, unlinked inversion coupling, unlinked idempotent coupling, address decoder, read/write logic, and write recovery faults.

- *March 10NJ*—March 10N with Johnson Counter background. This algorithm detects the same failure modes as the March10N algorithm. It also includes detection of bit-line coupling faults.

Diagnostics

Analysis of the interconnect between the ASIC and each of the memory pins within a cluster is achieved by a *stop-on-nth-error* option. This option supports an approach that tests the memory at-speed up to and including the first failed memory location. The failing address and data values are then shifted out to an external device that captures and stores this information.

The test is then repeated with the first failure ignored, stopping on the next failure and shifting out the failure data. The testing continues until all failure data is registered. Diagnostic software can be used to analyze the data and to produce reports that identify suspected interconnect flaws or mem-ory-bit maps that display failing bits.

Design Flow

The Sub-Block flow described in Chapter 4, "Embedded Test Design Flow," is used to generate a memBIST-XT controller and memory interface although the sub-block to test (memory module itself) is not part of the host ASIC. The controller is considered part of the user core logic. There-fore, the Prepare Logic Flow is used to insert the design objects resulting from the Sub-Block flow into the design. All other steps of the Prepare Logic Flow are performed to insert scan chains and generate a verification testbench. Finally, the Top-Level Flow is executed to obtain a final verifica-tion testbench that exercises the controller using the TAP interface. Test and diagnostic patterns are generated in SVF to provide links to system test.

Fault Insertion

Fault insertion (also known as *fault injection*) has been used for many years to evaluate the effective-ness of diagnostic software. Although it is possible to perform this evaluation using simulation, it is usually difficult because of the absence of proper tools and models.

A common way of performing this verification is to inject faults around components of a board using switches or other similar hardware. Usually, only a few carefully chosen faults can be inserted with this primitive approach. For each board revision, only a few boards can have the fault injection circuitry added. This method of checking the diagnostic software is expensive and ineffective, due to the limited number of faults that can be practically inserted versus the desired number of faults. Clearly, modern systems require better methods.

Boundary scan has several fundamental characteristics that make it very attractive as an infrastructure for fault insertion:

- Boundary scan provides easy access to board interconnect. The primary reason for using boundary scan to provide this access is to facilitate board manufacturing test. The same access mechanism can be used to inject faults for evaluation of diagnostic software.

- The boundary-scan infrastructure is designed to operate during normal system function. Because of this feature, the fault tolerance of the system can be evaluated.

- Boundary scan enables access to internal nodes of the integrated circuit for fault insertion. This access is becoming increasingly important because of the complexity of today's circuits.

Several levels of access to the potential fault sites (board interconnect and/or internal nodes) can be provided for the purpose of fault insertion. Access can be provided from the edge connector of each board, from an embedded test controller, or maintenance processor. The various access methods are described in detail in Chapter 8, "System Reuse of Embedded Test."

Fault Insertion Cell Types

Two types of fault insertion cells are discussed: correlated fault insertion cells and uncorrelated persistent fault insertion cells. Variations of these cells allow the insertion of faults at the pads or at nets in the core of the design.

Correlated Fault Insertion

For correlated fault insertion, the fault value (logic 1 or logic 0) is the same (*correlated*) for all fault insertion sites. The fault value is programmed from the TAP. An active fault insertion site is indicated by the presence of an appropriate fault flag in the update latch of a boundary-scan cell. These cells can be used if diagnosis of inserted faults using structural tests is not required (for example, logic BIST or board-interconnect test).

Uncorrelated Persistent Fault Insertion

For uncorrelated persistent fault insertion, different fault injection sites can have different fault values. Also, the fault remains in place in spite of new data being shifted through the boundary-scan register to perform structural tests.

Each potential fault site requires two update latches to provide a fault flag and a logic 0 or logic 1 fault value at each fault insertion site.

These cells occupy more silicon area but allow the use of boundary scan to inject faults *and* to diagnose them in the system.

Cell Design

This section describes the design of cells used to inject correlated faults or persistent faults at the pads or in the core logic.

Cells for Correlated Fault Insertion

Figure 7-6 on page 185 shows an output boundary scan cell augmented to implement correlated fault insertion.

- The *SelectJtagOutput* signal normally controls the multiplexer of the boundary scan cell that intercepts the functional data.

- The *FaultEnable* signal indicates whether the fault insertion mode is active.

- The *FaultPolarity* signal indicates whether a stuck-at 1 or stuck-at 0 fault is injected.

A fault is injected at a specific site (input or output pad) when the data contained in the update latch contains the value 1. If multiple faults are active, these faults exhibit the same polarity. Similar circuitry is used to inject faults at input and bidirectional pads. For bidirectional cells, a *FaultDirection* signal is required to inject the fault on the input (*FaultDirection* = 0) or output (*FaultDirection* = 1) path.

Figure 7-6 Cell for Correlated Fault Insertion into an Output

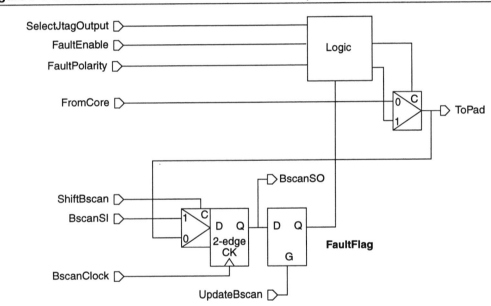

The signals *FaultPolarity, FaultEnable,* and *FaultDirection* are driven from the instruction register or a special data register embedded in the TAP. Use of a special data register is recommended and is the default assumed by the tools.

The cells for insertion of correlated faults require the addition of a small number of gates to the basic boundary-scan logic. However, these types of cells can only be used to inject faults in the system. These cells do not support the usage of structural tests such as an interconnect test or logic BIST to diagnose injected faults. If diagnosis with structural testing is attempted, the fault is removed before the structural tests can detect it. If structural tests are used to diagnose injected faults, cells designed for the insertion of uncorrelated persistent faults in the design must be used.

Cells for Uncorrelated Persistent Fault Insertion

To inject multiple faults that cannot be correlated onto a single chip, a memory element must be added to the boundary-scan cell. The polarity of the fault at a site can be independent of the one at other sites. In this case, a *FaultSetup* signal that replaces the *FaultPolarity* signal is required.

The additional latch stores the fault flag that indicates when a fault is injected. The *FaultFlag* signal is active high. If both the *FaultEnable* and *FaultFlag* signals are high, a fault is inserted. The value of the fault is determined by the existing boundary-scan update latch.

Figure 7-7 on page 186 is a schematic of the cell for *uncorrelated persistent fault insertion* at an output pad. The faults inserted with the cell *persist*; that is, these faults are not removed when structural tests (such as interconnect test or logic BIST test) are used to locate a fault.

Figure 7-7 Cell for Uncorrelated Persistent Fault Insertion into an Output

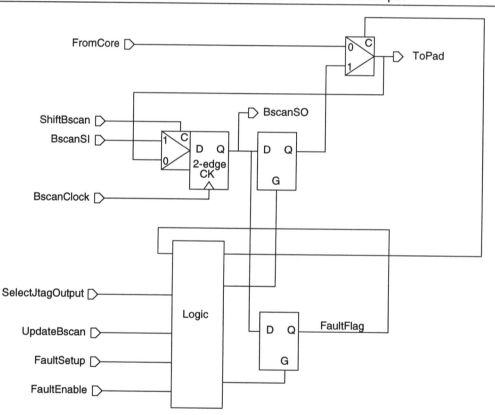

This persistence is an essential feature when software diagnostics are based on structural tests. In addition to the locking mechanism of the memory elements updated with the *UpdateBscan* signal, the control signals *FaultEnable*, *FaultSetup*, and *FaultDirection* must be sourced by a separate data register included in the TAP instead of the instruction register (IR). The separate data register cannot be used by system diagnostics software.

Two loads of the boundary-scan register, in addition to the ones required to access the instruction register and the special internal data register, are now required to inject faults. One load is used to set fault flags, and the second one to set fault values. The *FaultSetup* signal determines which latch is updated during a boundary-scan register load.

For bidirectional pads, a fault can be inserted in the input or output direction by using the *FaultDirection* signal described earlier.

Internal Fault Insertion Cells

Faults can be injected internal to the chip (in the core area) using essentially the same cells as those used to inject faults at the peripheral pads. The only difference is that the multiplexer intercepting the functional signal is instantiated manually in the core area.

Design Flow

Insertion of pin faults is completely automatic. Pins requiring the fault insertion capability need to be assigned the appropriate boundary-scan cell type during the Top Flow described in Chapter 4, "Embedded Test Design Flow. It is assumed that most faults are pin faults.

Internal boundary-scan cells are used to inject faults into the user core logic. The insertion of faults into the core of the design requires manually adding a multiplexer for each net that requires fault injection and connecting the control signal and test input to the corresponding internal boundary-scan cell at the top level of the design. This operation is performed before following the Core Flow described in Chapter 4.

References

[1] "IEEE Standard Test Access Port and Boundary-Scan Architecture," IEEE Standard 1149.1a, October 1993.

[2] B. Nadeau-Dostie, J.F. Cote, H. Hulvershorn and S. Pateras, "An Embedded Technique for At-Speed Interconnect Testing," in *Proceedings of the 1999 International Test Conference*, Atlantic City, N.J.

[3] B. Nadeau-Dostie, H. Hulvershorn and S. Adham, "A New Hardware Fault Insertion Scheme for System Diagnostics Verification," in *Proceedings of the 1995 International Test Conference*, Washington DC, pp. 994-1002.

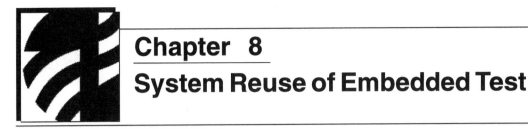

Chapter 8
System Reuse of Embedded Test

Embedded test capabilities contained within a chip can be reused during board and system testing as well as during field repair. The motivation for reusing embedded test lies in the unparalleled level of continuity between ASIC test and in-system test. Because the embedded test run in-system is identical to the test that was run when the device was manufactured, a device can be retested to the same level of fault coverage that was obtained during its manufacture without having to remove it from the board. This provides failure resolution to the ASIC level within a system, which is extremely useful when isolating a failure or trying to determine where in the manufacturing process a defect was introduced.

This chapter describes how to use LogicVision embedded test technologies to develop a complete board or system self-test capability.

Embedded Test

There are many benefits of using embedded test as part of the in-system test and maintenance strategy. Three primary concerns when trying to isolate faults within a system are fault coverage, test accuracy, and test execution time. With the runtime of an embedded test measured in milliseconds, stuck-at fault coverages in excess of 95 percent, and resolution to the ASIC or board net, system designers do not need to generate additional tests that attempt to isolate hardware defects. Instead, they can focus on the higher level "sanity checks" required before bringing a system on line.

The general system architecture that supports complete board or system self-test is illustrated in Figure 8-1 on page 190. This architecture has the following characteristics:

- *Most if not all* of the devices on a board contain at least IEEE 1149.1 boundary scan and, preferably, embedded test capabilities. Devices without at least boundary scan are very difficult to test and seriously impact the board's fault coverage.

- Each device that contains boundary scan has its 1149.1 TAP connected to a board-level 1149.1 test bus.

- Each board's 1149.1 test bus is connected to the backplane through a hierarchical 1149.1-based interface.

- One of the boards within the system contains some type of embedded test control mechanism. This control mechanism can be a dedicated device or a processor.

- The embedded test control mechanism is accessed by the operating system or an external source. This external source can be as simple as a PC or laptop PC.

Figure 8-1 Hierarchical Embedded Test Architecture

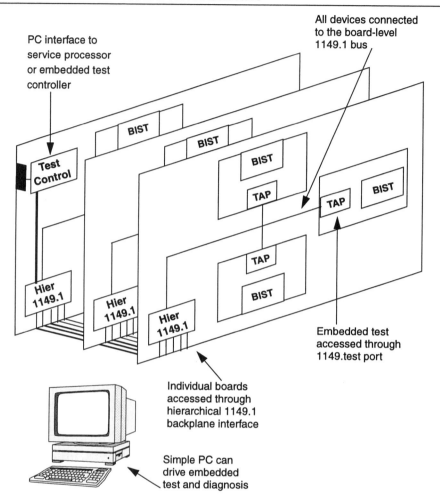

Adoption of an 1149.1-based test architecture requires that the decision-makers are familiar with embedded test and the 1149.1 standard, and are willing to depart from less effective but more famil-iar test methodologies. Although education and training can provide the necessary background for understanding the benefits of embedded test, the decision to use embedded test within a system usu-ally evolves slowly, beginning at the ASIC level, and ultimately moving to the system and network levels. Capabilities such as fault insertion or in-system reprogramming, where there is an immediate need to provide system-level 1149.1 access, are usually the catalysts for implementing a full system-level hierarchical test solution. The fact that embedded test can be exploited as part of this architec-ture is a positive side-effect but is not necessarily the primary reason for implementing the architecture.

There is no single system architecture that is *correct*, because the final implementation will be influenced by requirements that span multiple design and test groups. A good architecture meets the needs of everyone who requires access to the board's 1149.1 test bus. This includes access to embedded test, as well as other 1149.1 capabilities such as in-system programming, processor emulation, and fault insertion. It is critical that these requirements be researched in advance, with input provided by hardware designers, software designers, test engineers, and technicians. Without this research, it is possible to implement an architecture that is not fully used, or one that does not provide sufficient capabilities to be useful. It is as frustrating to implement a full-featured architecture that no one uses as it is to require access to capabilities that have been intentionally disabled.

In general, all implementations of a complete hierarchical test system require four basic functions:

- Test control—uses high-level commands to specify the test operations that are to be executed within the system. These commands, for example, **Run logic BIST U1** and **Run Memory BIST U1**, can be combined to form a **Run BIST U1** command, which can then be used as part of a board-level **Run BIST** command.

- Test pattern generation—translates test commands into a series of (1149.1) operations. The test pattern generator can either perform its function in real time or read pregenerated vectors from a file or memory.

- Hardware interface—applies the 1149.1 test operations to the system hardware. All data returned by the hardware is compared with the expected results. Any miscompares are either stored for later analysis or reported immediately to facilitate real-time failure analysis.

- Failure analysis—any failures detected during the sequencing of the test pattern must be translated into a meaningful diagnosis. Even tests such as logic BIST, which are localized to a single component, require analysis to ensure that the test failed because of a problem within the part rather than a failure of the 1149.1 bus or system clock. Failure analysis can be performed either in real time, or off line based on failure data stored by the hardware interface.

The decision about where these functions are implemented is dependent on the constraints imposed by the system. It is also likely that the partitioning of these functions will be dynamic, depending on the type of testing being conducted. The following example illustrates the partitioning of these test functions and how their distribution within the system can change. The example includes a first step used to detect the failure and a second step to diagnose its source.

1. A technician issues a **self-test** command to a system.

    - A technician uses a network interface to issue a self-test directive to a system. The system receives the command and broadcasts a **self-test** command to all boards within the system.

    - A board within the system receives the **self-test** command and breaks it down into the tasks **Run logic BIST U1**, **Run logic BIST U2**, and **Run memory BIST U2**.

    - For each task, the processor on the circuit card implements the test pattern generation function and reads the required sequence of board-level 1149.1 instruction and data register operations, including the expected response data, from a memory on the board.

    - Each 1149.1 operation is applied to the circuit card by the on-card processor. The response data is compared with the expected response and all failure information stored in an EEPROM on the board. The occurrence of a failure is reported to the technician over the network.

    Upon being notified of a failure within the system, the technician extracts the failure information from the EEPROM on the failing board. Based on the information within the memory, failure analysis run on the technician's workstation determines that logic BIST of U1 detected a failure within the device logic.

2. The technician wants to diagnose the logic BIST failure from a remote terminal.

- The technician generates a set of ASIC-level Serial Vector Format (SVF) files required to determine which logic BIST vector detected the failure. The ASIC-level pattern is modified to accommodate the position of the ASIC on the target board's scan chain.

- The board-level SVF file is read by the test pattern generator that is running on the technician's workstation. The pattern is appended with network and board-level addressing information, and the sequence of scan operations is transmitted to the system across the network.

- The target system applies the 1149.1 data to the addressed board using a dedicated backplane test bus. All response data from the target board is returned across the network for analysis on the workstation.

- The data returned by the system is compared against the expected data, concluding that vector 5 is detecting the failure. The process is repeated with the contents of the internal scan chain after running five logic BIST vectors returned for analysis. Analysis of the failing vector isolates the failure to flop *coreInst/myCore/D[0].Q*.

In this example, the infrastructure that was implemented supports the ability to isolate a failure to a failing flop within an ASIC while controlling the testing from a remote workstation. The actual infrastructure required for this access level was not overly complex and could be architected to use a combination of functional and dedicated communication paths. The following sections elaborate on the various board and system test access implementations.

Board and System Embedded Test Primer

When considering the topic of board and system embedded test, the first challenge is to define precisely what the *system* is. To ASIC designers, a system is the collection of modules required to implement the chip's function. To board designers, a system is a collection of ASICs, discrete components, a microprocessor, and firmware that controls all activity on the board. A collection of boards within a common chassis can also be considered a system, even if each board, or group of boards, within the chassis performs totally independent functions. The final definition of a system is a collection of software components, interconnected by a network, that perform functions resulting from stimuli received from both electrical and mechanical peripherals. Ultimately, the last definition is the most accurate, because any electronic system must perform its function flawlessly within the environment for which it was designed. Therefore, what does one test to ensure that the final system will perform its function?

The *divide and conquer* strategy is based on the assumption that an entire system functions properly if each component within that system performs its function correctly and is properly connected to other components. This approach is intuitive from a diagnostic perspective. For example, if two computers on a network cannot communicate with each other, the problem originates from one of the computers or within the connectivity between the two. If the manufacturer of a computer can guarantee that the computer is operating properly, and can properly control its network interface, then the problem is with the connectivity between the computers. Conversely, if the test of one of the computers fails, the source of the problem has been isolated and can be corrected.

If the same approach is applied across all levels of a system's hierarchy, an electronic chassis should be tested by ensuring that all boards within that chassis, and the interfaces between the boards, are functioning properly. Continuing further down the hierarchy, board functionality can be verified by ensuring that all devices on the board are functional and properly interconnected. Therefore, the first step of proving that a complex system is functional occurs at the device level.

What is necessary to prove that a device is functioning properly? Two types of problems can arise: design problems or problems related to defects. Although embedded test primarily detects and diagnoses manufacturing defects, some design defects can be addressed as well. For example, a clock skew problem between ASICs on a board can be diagnosed using *jtag-XLi*. Clock jitter can also be measured using *pllBIST*. Another example is to use LogicVision's fault insertion capability to insert faults in the system to verify that the device works as expected.

The basic rule of testing is that, in order to detect a defect, one must first apply a stimulus that triggers the defect and the effect of the defect must be observable. As the evolution of ASIC test demonstrates, it is no longer possible to attain sufficient fault coverage by applying a stimulus strictly to the I/O of a device (assuming access to the device on the board is possible). Techniques such as embedded test are used to apply test stimuli and monitor circuit response deep within a device at system clock rates. The objective of embedded test is to exercise all logic and data paths within the device at functional speeds, proving that the logic and interconnects within the device can operate correctly at functional speeds. Any defect that affects the functional behavior of the device is detected when the embedded test is run. The fundamental question embedded test answers is whether a device is defect-free and will therefore operate as it was designed. This is critical in determining whether a failure is the result of a defect or is a design marginality issue.

After the device has been placed on the board, the problem of proving that a device is still defect-free remains, because defects can be introduced anywhere in a manufacturing process. A device can also "wear out" prematurely because of defects such as gate-oxide shorts. Finally, the device behavior might depend on environmental conditions (for example, supply voltage, noise, temperature, and so forth) that cannot be reproduced at device manufacturing time. A difference between ASIC-level testing and board-level testing is that the erroneous behavior of a device can be attributable to factors external to the device, including defects within other ASICs or board-level faults such as bad solder joints. To diagnose whether a device is the source of a problem, one must either be able to rerun the tests performed at the device level, or attempt to do what the designer of the device could *not* do: namely verify that the device is defect-free by using functional tests. By leveraging the embedded test capabilities embedded within all devices on the board, isolation of ASIC-level defects is instantaneous. Board-level tests need only target those defects not covered by the embedded ASIC-level tests—specifically the interconnects on the board. Similar to the strategy employed at the ASIC level, a board that is proven to be defect-free will function as designed.

Every additional level of hierarchy within the test process builds upon the tests and infrastructure put into place by the previous level. By encapsulating the test function at every stage of hierarchy, there is no degradation in fault coverage and fault isolation capabilities within the final system. It is as easy to diagnose an ASIC-level defect from across a network as it is to diagnose an ASIC-level defect at the board level.

Benefits of Using Embedded Test

The availability of embedded test at the board and system levels provides access to capabilities that can be reused throughout product development and deployment. By exploiting these capabilities, test development time is dramatically reduced while increasing the quality of ASICs, boards and systems. The following is a partial list of benefits that can be derived at various stages of the product cycle.

Benefits for Design Engineering

When the first run of boards is manufactured, minimal testing of the boards is performed because no debugged ATE or functional tests are available. The design engineer is responsible for initial test of the board, which includes finding any manufacturing *and* design defects. There are no golden boards yet.

Embedded test provides the following benefits:

● Higher quality of ASIC. There is a high level of confidence that the part on the board is working (as mentioned previously, some design problems can also be identified), reducing the amount of time spent trying to debug ASIC-level problems. ASIC-level functional tests do not guarantee working parts. Scan testing of an ASIC is an improvement over functional testing but cannot catch speed-related faults (assuming again that those tests are available and applicable). A functional test of the board provides an at-speed test of the device, but the coverage and ability to diagnose which device is causing the problem is reduced.

● Ability to instantly retest an ASIC. If an ASIC is suspect, the embedded tests used at the ASIC level can be rerun to verify whether a defect has been introduced. Without embedded test, the engineer would remove and replace the part, hoping that the ASIC was in fact defective. Turnaround time for replacing a device is usually greater than one day; whether the part was actually defective will not be known until failure analysis on the removed part has been completed. There is no guarantee that the removal of the part did not introduce a defect; the part was defective in the first place; the functional test used to test the ASIC exercises the defect or that failure analysis can or will be performed.

● A passing embedded test indicates that the ASIC is still defect-free, and that the problem is elsewhere on the board, or that there is a design (marginality) issue. In the remove-and-replace strategy used if embedded test is unavailable, the board may start working again after a new part is installed. This could be a result of the new part having different operating characteristics than the original. Therefore, the design engineer will not be aware of a potential marginality issue—at least not until the boards go into volume production.

● The operation of the embedded test controllers has been simulated. If an embedded test fails at the board level, the exact test scenario can be simulated at the ASIC, or board level. This is because the Verify tool used to generate the simulation testbench is the same tool used to generate the board and system-level tests. Also, because the test used at the board level has been simulated, the probability of the test passing on a defect-free board is very high. Board-level functional tests are rarely simulated.

● The board can be verified in the absence of firmware. The first run of boards typically must be debugged before firmware is available. (In reality, the firmware groups need access to a defect-free board in order to debug their firmware.) This parallelism allows the hardware and software groups to perform their verification work independently.

● Failure resolution, in milliseconds, to the device or board net. Pulling out logic analyzers for diagnosis is time consuming, and unnecessary. If a board is failing, all embedded tests can be run in seconds using only the lab PC and an IEEE 1149.1 interface. If the test fails, the design engineer has resolution to the ASIC or board net. If the test passes, more in-depth diagnosis can be performed.

Benefits for Manufacturing and Failure Analysis

Traditionally, test programs are developed outside of the design environment, requiring the test engineer to develop tests based on design specifications without access to the simulation environment. To work around this problem, test engineers typically use the suite of system-level tests developed by the designer engineers for design verification. As a result, test engineers perform design verification on every system rather than running tests that are targeted at finding manufacturing defects. This practice is very common, time-consuming, and unnecessary.

Because there is no concurrency in the testing process, it is likely that system test will find defects that are not caught by board and ATE testing. This lack of concurrency causes a situation in which board and ATE tests must be updated constantly to target faults detected at the system level.

Embedded test provides the following benefits:

- Embedded tests are *defect* oriented, so the tests developed by the design engineer can be reused directly. There is no additional test development necessary and the design engineer has already done the work of debugging the tests.

- Because embedded test provides a common suite of tests that can be used throughout the test process, ATE, functional, and system test groups do not have to generate tests to target defects covered by embedded test. Because ATE and board test can run the embedded test, it is guaranteed that the embedded test will pass within the system, unless a defect has been introduced within the manufacturing process. Also, if an embedded test fails in-system, the failing card can be sent directly for repair without being sent back to board test to determine the cause of failure.

- If temperature cycling or burn-in is used, embedded test provides maximum node-toggling and failure isolation capabilities required for effective environmental testing. Because the testing is contained within the system (minimal connection required for the tester), a failure can be attributed to the system rather than the test equipment conducting the test. Also, the test strategy used to burn in a board within a system is *identical* to the test strategy to burn in spares. Typically, two separate test processes for full systems and spares must be developed.

Benefits for Software Groups

When a system fails, engineers typically assume a software problem. Software is ultimately responsible for the correct operation of a system, including the problem detection, recovery, and fault isolation. More than half of the system software is oriented at failure detection, recovery, and isolation, yet very little testing of the system in the presence of faults is performed.

Embedded test provides the following benefits:

- If a system malfunctions, running embedded test quickly determines whether a hardware defect is the cause. Software developers can spend time debugging software problems rather than trying to debug problems caused by defective hardware.

- Boundary scan fault insertion allows software groups to verify that faults within a system actually invoke the correct recovery and diagnosis routines. Software typically relies on hardware to detect failures, and based on flags set by hardware, will invoke the appropriate routines. Fault insertion verifies that the hardware is raising the correct flags and that the system will actually recover. Although the behavior of the system under fault conditions is considered during system design, some side-effects of the fault might not be anticipated.

- Diagnosing hardware defects from within software is nearly impossible. By providing access to hardware-based embedded test capabilities, software has access to the fault detection capabilities that are targeted at finding hardware defects.

- Time constraints can restrict the amount of testing software can perform before the board or system must become operational. This typically limits the testing to a basic memory test and checks to ensure that the processor can read and write selected registers within each ASIC. This approach increases the risk of a board failing when it is in service because a defect is not detected until the functional operation of the board exercises the defect. The execution time of embedded test can be exploited as part of a board or system's power-up testing strategy.

General Embedded Test Controller Architecture

This section describes the general architecture common to all LogicVision embedded test controllers. Only the aspects of the architecture that pertain to the hierarchical access and reuse of the embedded test controllers are described.

The TAP/Embedded Test Controller Interface

The interface between the IEEE 1149.1 TAP and any LogicVision embedded test controller is shown in Figure 8-2 on page 198. The interface is made up of a scan interface and direct controls. A description of the signals that make up the interface is provided below:

- *BIST_SI*—input to the embedded test controller that serves as the scan input for accessing the controller's internal registers.

- *BIST_SO*—output from the embedded test controller that serves as the scan output during retrieval of test result and diagnostic data.

- *BIST_SETUP*—two-bit input bus to the embedded test controller that specifies the run and setup modes for the controller.

- *BIST_ENABLE*—input to the embedded test controller that enables the controller for either scanning or execution depending on the value of the *BIST_SETUP* bus.

- *BIST_DONE*—output from the embedded test controller that indicates when the controller has completed execution.

- *BIST_GO*—output from the embedded test controller that provides the pass/fail result. This output is normally high throughout execution of the embedded test. If a failure is detected sometime during the embedded test, this signal is driven low either at the end of the embedded test, or for some embedded test controller types (such as memory BIST), as soon as the failure is detected.

- *TCK*—input to the embedded test controller that is driven by the TAP test clock.

The *BIST_SI*, *BIST_SO*, *BIST_SETUP*, *BIST_ENABLE*, and *TCK* signals make up the embedded test controller's scan interface. All LogicVision embedded test controllers have an internal scan chain that is referred to as the *LONG_SETUP* register. This register contains both the initialization and result bits. It is possible to access only the result bits. This subsection of the *LONG_SETUP* register is referred to as the *SHORT_SETUP* register. Details on these two registers are provided in the section "Embedded Test Controller Registers" on page 198.

The 2-bit *BIST_SETUP* bus is used to select which of the two registers to scan. The complete function of the BIST_SETUP bus is described in the following table:

BIST_SETUP[1:0]	Selected Mode
1 1	Short Setup
1 0	Long Setup
0 1	Default Run
0 0	Normal Run

In addition to choosing which register to scan, the *BIST_SETUP* bus determines in what mode the embedded test controller runs. In Default mode, the embedded test controller ignores any scanned in initialization data and runs based on hardcoded default initialization values. In Normal mode, the embedded test controller runs based on the scanned-in initialization values.

All scan operations between the TAP and embedded test controller must occur at *TCK* rates. Because all LogicVision embedded test controllers provide at-speed test capability, they must run off of the system clock (*BIST_CLK*). Therefore, to support *TCK*-based scanning, each embedded test controller contains an asynchronous interface that permits its *LONG_SETUP* and *SHORT_SETUP* registers to be scanned at the *TCK* rate while being clocked by the system clock.

To allow scanning to occur, or to kick off test execution, the embedded test controller must be enabled using the *BIST_ENABLE* signal.

As shown in Figure 8-2 on page 198, the *BIST_SETUP* and *BIST_ENABLE* signals originate from TAP Instruction register bits. Therefore, all embedded test controller scan operations consist of a combination of TAP Instruction and *LONG_SETUP* or *SHORT_SETUP* register scans.

The remainder of the TAP/embedded test controller interface consists of the two result signals, *BIST_DONE* and *BIST_GO*, which are routed from the embedded test controller to TAP Status register bits. Examination of the values driven on these signals is therefore accomplished by scanning out the Status register.

All 1149.1 TAP data sequences needed to initialize, execute, and extract results from an embedded test controller can be generated automatically using LogicVision's Verify automation tools. For example, the logicbistVerify tool allows the user to specify various runtime options. Based on this input, the tool generates an SVF data file that contains all of the TAP sequences needed to initialize a given logicBIST controller with the desired options, run the controller to completion, and extract the final pass/fail result. As described in the section "Pattern Resequencing" on page 202, this SVF file can be resequenced and applied at the board or system level.

Figure 8-2 Interface Between Embedded Test Controller and IEEE 1149.1

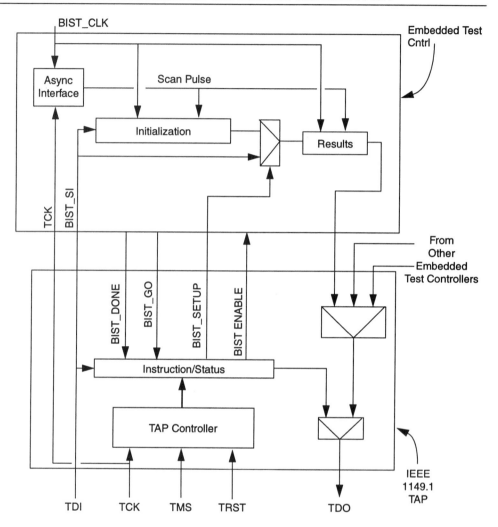

NOTE: *BIST_SI* and *BIST_SETUP* connect to *all* embedded test controllers. There is one
BISTENABLE signal per embedded test controller.

Embedded Test Controller Registers

All LogicVision embedded test controllers are configurable using two 1149.1 accessible data registers. These data registers, the *Long Setup* and *Short Setup* registers, are used to specify runtime
parameters, and to extract information from the controllers if an embedded test fails. The Long
Setup register allows access to all configuration and failure information registers within the controller and is intended for detailed failure analysis. If the detail provided by the Long Setup register is
not required, the embedded test controller can be accessed using the Short Setup register. Physically,
the Short Setup register comprises selected bits of the Long Setup register. The objective of the Short
Setup register is to allow the embedded test controller to be initialized and a go/no-go status to be
read using a minimum length data register. For some embedded test controller types, depending on
the options chosen, the Long Setup and Short Setup registers might be identical.

Information about the function of bits within the Long Setup and Short Setup registers is contained within the Hierarchical Scan Description Language (HSDL) file generated by LogicVision's automation tools. Although using the Verify tool to generate the test pattern for an embedded test is strongly recommended, the information contained within the HSDL file is useful when performing failure analysis. A complete description of the HSDL syntax can be downloaded from the following Web site:

```
http://support.asset-intertech.com/hsdl.htm#download
```

The following sections of an HSDL file were generated for a test chip implementing logic BIST. Only the sections of the HSDL description relevant to the Setup registers are presented.

REGISTER ACCESS

The *REGISTER_ACCESS* attribute describes the length of a data register and the mnemonic of the instruction used to select the register. In this example, the Long Setup and Short Setup registers are both declared as being 117 bits long and have a bit range of [116:0], as shown in Figure 8-3. Bit 0 of any register defined by this attribute is always the bit closest to *TDO*; that is, the first bit shifted out of the device. The Short Setup register is selected when the instruction register contains the value defined by the *LBIST_SHORT_SETUP* instruction defined within the *INSTRUCTION_OPCODE* attribute of the HSDL file. Similarly, the *LBIST_LONG_SETUP* instruction is used to select the Long Setup register.

```
attribute REGISTER_ACCESS of LVBIST1A: entity is
    "LBIST_SHORT_SETUP_REG[117](LBIST_SHORT_SETUP)," &
    "LBIST_LONG_SETUP_REG[117](LBIST_LONG_SETUP) ";
```

Figure 8-3 Short Setup and Long Setup Registers

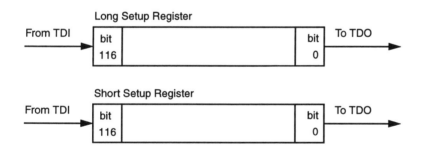

REGISTER COMPOSITION

The Short Setup register physically comprises bits within the Long Setup register. The *REGISTER_COMPOSITION* attribute defines which bits of the Long Setup register are selected when the Short Setup register is active. In this example, the Short Setup and Long Setup registers are identical.

```
attribute REGISTER_COMPOSITION of LVBIST1A: entity is
 "LBIST_SHORT_SETUP_REG (LBIST_LONG_SETUP_REG[116,0])";
```

For illustrative purposes, if the Short Setup register is two bits long and consists of the first and last bit of the Long Setup register, the HSDL attribute is the following:

```
attribute REGISTER_COMPOSITION of LVBIST1A: entity is
 "LBIST_SHORT_SETUP_REG (LBIST_LONG_SETUP_REG[116]," &
 "LBIST_LONG_SETUP_REG[0] )" ;
```

The bits are defined in order from *TDI* to *TDO*. In the preceding example, bit 116 is the bit closest to *TDI* and bit 0 is closest to *TDO*.

BUS COMPOSITION

The *BUS_COMPOSITION* attribute is used to identify the function of the bits within a data register. A meaningful name is assigned to a segment of the data register and the position of this subregister is identified. The logicBIST controller for the example defined above contains the following register segments.

```
attribute BUS_COMPOSITION of LVBIST1A: entity is
"LBIST_PRPG[24](LBIST_LONG_SETUP_REG[25,48])," &
"LBIST_MISR[24](LBIST_LONG_SETUP_REG[1,24])," &
"LBIST_VECTOR_CNT[16](LBIST_LONG_SETUP_REG[83,98])," &
"LBIST_BYTE_CNT_MAX[8](LBIST_LONG_SETUP_REG[109,116])";
```

The first *BUS_COMPOSITION* statement assigns the name *LBIST_PRPG* to the register segment of the logicBIST controller's Long Setup chain that corresponds to the logic BIST PRPG. This register segment is 24 bits long and comprises the Long Setup chain bits 25 through 48. This allows the specification of a PRPG value to be mapped to the correct bits of the Long Setup chain. The bit ordering specified by this attribute is critical. The register segment is specified from most significant bit (MSB) to least significant bit (LSB). For example, the MSB of the PRPG value is bit 25, and the LSB of the PRPG value is bit 48 (see Figure 8-4).

Figure 8-4 Bit Ordering of Subregisters

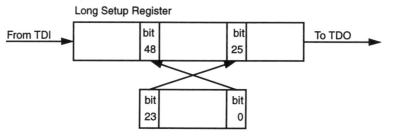

Logic BIST PRPG Register

Pattern Generation

Although the HSDL file describes the structure of the registers within a controller, there is no information regarding the sequence of instruction and data register scan operations that make up the algorithm required to initialize and run the embedded test controller. The algorithm for running an embedded test is generated by the embedded test controller's Verify tool. Using the Verify tool to generate an SVF-formatted pattern that can be retargeted for use at the board and system levels is strongly recommended.

This strategy ensures that the pattern applied to the controller is correct by construction, reducing the time required to debug the tests. The Verify tool can also be used to generate a simulation testbench to validate the test before being applied at the board and system levels.

Board-Level Test Access

Board-level access to LogicVision's technology is provided through a board-level 1149.1 bus. All 1149.1-compliant devices are connected to the bus with *TMS*, *TCK*, and *TRST* connected in parallel and the *TDI* and *TDO* pins of all devices chained together to form a serial scan path on the board.

Test control, test pattern generation, hardware interface, and failure analysis functions are all performed off card either by conventional ATE or an 1149.1-based test system such as those available from ASSET InterTech, Corelis, Intellitech, or Jtag Technologies.

Hardware Requirements

In addition to providing testpoint access for ATE testing, a 10-pin connector (5-signal, 5-ground) is typically used to interface to the 1149.1 bus. This connector provides the interface to drive *TRST*, *TCK*, *TMS*, and TDI, and monitor *TDO*. This method of providing test access is the least expensive in terms of hardware overhead but is the most restrictive because the connector might not be accessible within a fully populated system. In such a situation, the card must be removed from the system to perform any 1149.1 test operation.

In addition to the connector, a buffer on the *TCK* signal is recommended to prevent false clocking of the 1149.1 data. Pull-ups on the *TRST**, *TMS*, *TCK*, and *TDI* signals are required to maintain safe values on the bus when no test equipment is connected to the board. With the possible exception of the *TCK* signal, whose termination is not specified within the 1149.1 standard, the pull-ups on the board match the termination required within the device. Additional buffering and bus termination requirements for the 1149.1 bus signals are usually specified within the test system's hardware interface documentation.

More complex circuit cards may require multiple 1149.1 buses to support functions such as processor emulation, or to provide access to optional daughter cards. If possible, the presence of additional scan chains should be avoided as it adds complexity to the external test system, and complicates the generation of board-level 1149.1 test patterns. Using embedded test significantly reduces the need for multiple scan chains because the entire board can be tested with a minimum amount of information to be transferred through the bus. If multiple scan chains are unavoidable, it is recommended that any multi-scan chain board be reconfigurable to a single scan-chain board through the use of jumpers or scan path linkers.

Jumpers for Scan-Chain Configuration

The simplest way to reconfigure the 1149.1 bus on a board is by using jumpers. This method hard wires the scan chain, allowing test patterns to be generated for a specific board configuration. If a daughter card is added to the board, the jumpers on the board are set to include the new card on the bus, and the test patterns regenerated for use with the new scan-chain configuration. This strategy requires that several sets of test patterns be generated and maintained.

Scan Path Linkers

Scan path linkers are 1149.1 devices that allow the selection of one or more secondary scan paths on the board. The scan path linker is connected to the board's primary scan path so that its instruction and data registers are always accessible. By loading the instruction and data registers of the scan path linker, any combination of secondary scan paths on the board can be selected. Once activated, the secondary paths are part of the board-level scan chain, and must be considered when generating a board-level test. The scan path linker can also concatenate secondary scan paths, providing a single scan chain required for interconnect ATPG.

Because the scan path linker is configured using the 1149.1 protocol, the 1149.1 operations required to configure the scan path linker can be appended to other 1149.1 test sequences. This allows a test sequence to be generated that is independent of changes to the board configuration. For example, if a device resides on a secondary scan path that is not affected by board variables such as daughter cards, the combination of 1149.1 operations to configure the scan path linker and those required to conduct the testing of the device remains constant. This strategy helps minimize the quantity of test vectors that must be maintained.

Test Control

All functions associated with the 1149.1 bus are provided through either external ATE or an 1149.1 test system. The external test system is responsible for controlling the sequence of tests to be run, generating the test vectors associated with each test, applying the test to the hardware, and performing any failure analysis based on the actual data returned by the hardware.

The following companies distribute products that perform the test control function:

- ASSET InterTech—provides several ATPG functions and supports the application of user-generated patterns. Hardware interface functions are accessible from within supplied tools, within a test macro language, and through C++ interface functions.

- Intellitech—provides several ATPG functions and supports the application of user-generated patterns. Hardware functions are accessible from within supplied tools and from within a high-level test language.

- Corelis—provides several ATPG functions, and supports the development and application of user-defined patterns.

- Teradyne—Victory software provides ATPG for board interconnects. Both ASSET and Intellitech use Victory as their interconnect test ATPG.

Pattern Resequencing

The LogicVision Verify tools are used to generate the test sequences required to execute an embedded test. These test sequences are based on parameters specified by the user within the Verify tool configuration file, allowing the user to manipulate controller runtime parameters, either for design verification or failure diagnosis purposes. Encoded within the Verify tool is the algorithm that initializes the controller, calculates the test runtime, and shifts out (and compares) the test results. This eliminates errors that can inadvertently be introduced during the test generation process, because all knowledge of the hardware and the test algorithm is encapsulated within one tool. The Verify tool can also be used to generate a simulation testbench, allowing any test generated for use at the system level to be simulated before being run. If the test is to be reapplied at the board level, the Verify tool can be instructed to generate an SVF-formatted file. (For more information about SVF, refer to the Web site http://www.asset-intertech.com/svf.htm.)

The SVF file contains the sequence of 1149.1 instruction and data register scan operations that must be applied to a target device. The generated pattern does not account for the position of the device on the board-level scan chain. Every instruction and data register scan operation must be modified to account for the instruction and data registers of devices in front of and behind the device for which the pattern was generated. Some test systems allow the user to associate an SVF file with a device on the board; these test systems add the required instruction and data register padding based on the board's test bus configuration.

If a simpler test system is used, such as an SVF to PC parallel port translator, the SVF pattern can be manually modified using the SVF header and trailer directives:

```
HIR <length> TDI(<hex pattern>) TDO(<hex pattern>) MASK(<hex pattern>);
TIR <length> TDI(<hex pattern>) TDO(<hex pattern>) MASK(<hex pattern>);
HDR <length> TDI(<hex pattern>) TDO(<hex pattern>) MASK(<hex pattern>);
TDR <length> TDI(<hex pattern>) TDO(<hex pattern>) MASK(<hex pattern>);
```

The *<length>* specifies the length in bits of the header (HIR, HDR) or trailer (TIR,TDR) pattern. A different header can be used for Instruction Register (IR) access and Data Register (DR). *TDI(<hex pattern>)* specifies the data pattern expressed in hexadecimal to be applied to the TDI pin of the board. *TDO(<hex pattern>)* specifies the pattern, expressed in hexadecimal, expected on the TDO pin on the board. *MASK(<hex pattern>)* is used to specify which bits within the specified TDO pattern are to be compared with the actual response data. A "1" in the mask indicates that the corresponding bit specified within the TDO field is to be compared with the actual response data, a "0" represents a don't care or "X" value.

The above paragraphs assume that the device for which the pattern is retargeted has a fixed position on the board scan chain. This is not the case when scan path linkers are used. The ASIC-level pattern must be padded in accordance with the selected board-level scan path. If the configuration of the board changes, then the device's SVF pattern must be retargeted with a new set of headers and trailers. Many of the 1149.1 test systems understand the behavior of the scan path linkers and automatically apply the initialization sequence to the scan path linker and adjust the padding of the device SVF in accordance with the active board-level scan chain.

Test Pattern Generation

Beyond exercising ASIC embedded test capabilities, additional board-level tests must be applied. These include board-level interconnect tests, ASI tests, board-level memory tests, and cluster logic tests. The board interconnect test uses the boundary scan capabilities within each device to exercise the nets on board. Two sets of interconnect tests must be generated. The first interconnect test exercises all board-level nets, including those routed to the card edge connector, provided that the test system can drive and monitor signals at the card edge. The second pattern is restricted to testing only those nets contained within the card, requiring no external stimulus. This second pattern is required for in-system test access as described in the next section.

Several boundary scan-based ATPG packages are available that can generate a boundary-scan interconnect test based on information read from the board netlist, the BSDL files for each boundary scan-compliant device, and several constraints files. The interconnect test places all 1149.1-compliant devices into EXTEST and applies a set of boundary scan vectors to exercise the board-level nets.

The number of vectors required to fully test the board is calculated as the log of the number of board nets, which for a typical board is approximately 10 boundary scan vectors. The effectiveness of the test is a function of the number of board nets that connect to only 1149.1-compliant devices, and the grouping of the boundary scan cell enables within each device.

LogicVision's jtag-XLi at-speed boundary scan capability allows board nets to be tested at application speeds while retaining the diagnostic resolution provided by boundary scan. When generating the at-speed interconnect test, the alternate BSDL file generated by the top-level integration tool must be used in order to prevent the ATPG software from driving the clock and at-speed sync signals. The modified BSDL description declares all boundary scan cells associated with the sync and clock pins as internal cells. As a result these pins are treated as non-scannable for test generation purposes. An important assumption about the board design required by jtag-XLi is that a system clock must be present for all devices involved in the at-speed interconnect test. The pattern generated by the ATPG tool must therefore not affect the system clocks on the board.

Many aspects of a board design can greatly impact the test efficiency of a boundary scan interconnect test. As a guide, coverage that exceeds 50 percent is exceptional and coverage of 30 percent is probably typical. Issues that impact the board-level interconnect pattern generation include the following:

- Bad implementations of 1149.1 within ASICs. Board-level constraints are rarely considered when the ASIC designer is implementing boundary scan. A common mistake in the implementation of boundary scan is improper grouping of enables. Optimally, every device output should have a single boundary scan enable cell associated with it, however, in practice, most design engineers use a single boundary scan enable cell to control multiple outputs. If any of these outputs is connected to a net that cannot be controlled using boundary scan, then ALL signals controlled by that enable cell could potentially be omitted from the test.

- Another common mistake is not ensuring that every net can be controlled using boundary scan. This includes control over the output enables of non-boundary scan devices, and ensuring that once non-boundary scan parts are disabled, that at least one device connected to that net has an output drive capability. The best way to catch these pitfalls is to generate an interconnect test early in the ASIC design cycle and modify the boundary scan chain implementation to work around these issues. Alternatively, a board-level design rule checker must be run. However, such a commercial tool does not yet exist.

- Non-1149.1 components negatively impact the coverage obtainable using boundary scan techniques. The law of DFT surrounding controllability and observability are seldom applied at the board level, partially due to the cost or unavailability of boundary scan components. In some cases the presence of a non-boundary scannable part may be tolerable, such as a boundary scannable IO connected to the same net as a non-boundary scannable input. Such a combination still allows testing the net for shorts, however, it is impossible to test for opens between the two components. Most board-level ATPG tools do not have any component modeling capabilities and therefore cannot deduce how blocks of logic can be disabled or tested.

- The presence of memory on the board also affects coverage, because no memories implement boundary scan. *memBIST-XT* is an effective way of testing for both opens and shorts between ASICs and memories. Other approaches, including using boundary scan to perform memory accesses, have been successfully implemented, although the test application time for a complete memory test is significantly longer.

- Analog components. Similar to non-boundary scan components, analog components can negatively impact the coverage obtainable using boundary scan. The ATPG tool generating the interconnect test must support modeling for analog components to determine whether the component has an affect on the test. Examples of such components include series resistors, which act as wires for test generation purposes, and pull-up or pull-down resistors, which are considered "opens" during test generation. Although modeling allows a digital pattern to be generated and applied, the test cannot ascertain whether an analog component is the correct value. The 1149.4 standard addresses the testing of analog components by allowing an analog stimulus and resulting response to be routed to or from a pin on an 1149.4-compliant device.

- Clusters of logic. There are no commercially available board-level ATPG products that can perform vector generation to test combinatorial or sequential logic on a board. In addition, few systems support the serialization of parallel vectors using the boundary scan chain. A fundamental problem is that test patterns and device models that are used for test generation are generally developed for use at the ATE level and are tester specific. This information is not portable for use within an ATPG environment. Another issue pertaining to Cluster ATPG is that most board designers do not perform board-level simulations. Without the need for simulation, VHDL or Verilog netlists and device-level models are rarely developed or available.

In-System Card Test Access

The next phase in the testing process is to access the 1149.1 bus from within the system itself. Unlike the previous test stages, where the circuit card is fully accessible, on-card test access may be required within a fully populated system after all board hardware, including EMI shields, has been installed. The strategies at this testing stage vary greatly depending on the requirements and constraints being considered. These strategies range from extending the board-level 1149.1 bus across the system backplane to fully integrating the 1149.1 embedded test capabilities with system software.

When planning to use structural tests, it is important to consider that, when a test is being run, the device's characteristics are changed from its functional mode to a completely different test mode. For example, while logic BIST is being run, all outputs of the device that are equipped with boundary scan are tristated. Although systems are designed to accommodate devices being tristated (power-up and reset), the board designer must be aware that this mode exists when the device is being tested and must accommodate that mode by using pull-ups on the board. (Outputs that always drive the bus to a known state functionally cease to do so when placed in test mode.) Also, for all structural-based tests, including logic BIST, memory BIST, and boundary scan tests, the device's internal logic is left in a random state and might not be able to resume its functional operation mode. This side-effect of running an embedded test requires that the device, or even the entire board, is reset at the completion of all testing. Once again, these issues can be accommodated during board design.

Hardware Requirements

The hardware requirements to support on-card test access vary depending on the approach that is implemented. This section describes three of the most popular approaches.

1149.1 Card Edge Access

The 1149.1 card edge connector can still be used to access the diagnostic features of a board. This assumes that physical access to the 1149.1 card edge connector is possible, and that an external 1149.1 test system will be used to drive the test bus. This solution does not provide any hierarchy in the testing process, because the board and system software cannot leverage the availability of embedded test. This architecture is difficult to implement for backplane interconnects because multiple 1149.1 test systems, or a test system equipped with multiple 1149.1 bus controllers, are required.

It is worth noting that testing is conducted autonomously to the card's operation, which forces a card out of service. System messaging must be used to "shut down" the card on which testing is to be conducted.

1149.1 Backplane Extensions

Access to the board's 1149.1 bus can be provided by extending the board-level 1149.1 bus across the backplane as shown in Figure 8-5 on page 206. This is achieved by adding a 4- or 5-wire bus connected to all circuit cards within a system. Because of its multi-drop nature, boards that connect to the test bus are assigned a unique address that allows individual access of a board *dot1* bus. The added complexity of adding an addressing layer to the access of the board 1149.1 bus usually requires that a backplane test bus interface be used on each board. Using an 1149.1 backplane extension provides a single point of access for all cards within the system.

Figure 8-5 Board-Level Test Access Using an 1149.1 Backplane Extension

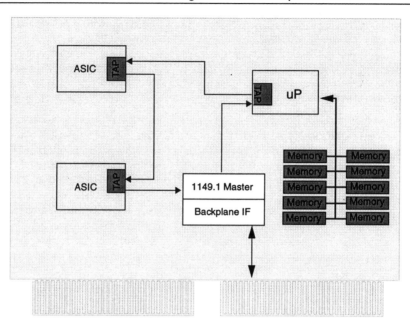

These devices are readily available commercially, and include the scanBridge components from National Semiconductor, and ASP devices from Texas Instruments, Inc. IEEE P1149.5 is another example of an 1149.1 backplane extension that is extensible to include test and maintenance functions other than only 1149.1 access. To date, implementations of P1149.5 have been proprietary and specific to the test and maintenance needs of a particular system.

On-Card Processor Access

To fully integrate embedded test with the card's diagnostic capabilities, testing must be controlled by a microprocessor, as shown in Figure 8-6 on page 207.

The hardware interface to the board is provided by an on card test bus master responsible for translating processor read and write accesses into the control and data streams on the 1149.1 bus. The test bus master can be implemented as a separate part on the board, integrated as part of an existing part on the board, or implemented within software using General Purpose I/O (GPIO) or Serial Peripheral Interface (SPI) capabilities of the processor itself.

If the on-card processor is to maintain control of the test bus, the board designer must ensure that none of the test modes exercised affects the operation of the processor. This applies not only to the processor itself, but also to any device that retrieves instructions and data for the processor. The designer must also provide communication to the processor to control the tests being executed. This communication path can also be used to provide debug access to the card if the on-card diagnostics detect a problem. Once again, the designer must ensure that this communication channel remains operative for all test modes, and in the presence of any faults that affect the board's operation.

Figure 8-6 Software Access to the 1149.1 Bus

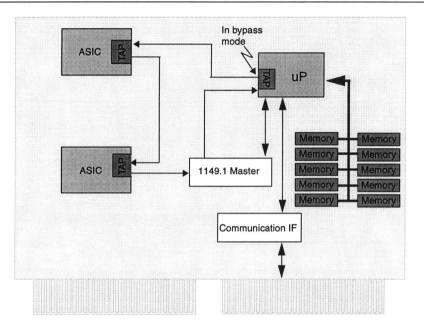

Because the pattern generation and resequencing operations are controlled by software, it is necessary for the software team to be involved early in the hardware design phase to provide 1149.1 functionality for hardware debug purposes. National Semiconductor makes available freeware that performs the functions required for processor-based 1149.1 access. (The ScanEase Web site is http://www.national.com/scan.)

Autonomous Embedded Test Controller

To solve the problems associated with processor-based embedded test, an embedded test controller can be used on the board to conduct all 1149.1-based testing, thereby minimizing the constraints associated with using the functional processor. The embedded test controller performs a pattern generation and/or resequencing function, typically using patterns or parameters stored in a dedicated PROM. The hardware components required to implement this architecture are shown in Figure 8-7 on page 208.

Figure 8-7 Using an Autonomous Embedded Test Controller

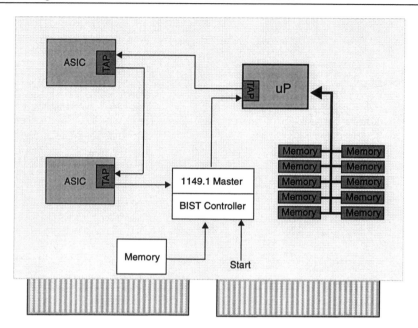

The method to start the embedded test controller must be addressed. For example, does a power-up reset automatically invoke the embedded test controller? Can tests be initiated by the card processor? From the backplane? Both? The answers to these questions define what interfaces must be provided to the embedded test controller. As a general practice, any test and maintenance capability is shared between the local processor and a backplane test and maintenance bus. This provides the most flexible means of accessing the embedded test capability, allowing tests to be initiated even if the processor becomes inoperative.

Test Control

Control of the testing function for systems employing 1149.1 test bus extensions is typically provided external to the system within which the testing is conducted. The setup for conducting the tests is similar to the setup that is required for testing using a card edge connector.

The pattern generation function is performed on a workstation, creating the sequence of instruction and data register scan operations required to conduct the test. These test sequences are bundled with addressing information and transmitted across the backplane using the protocol specific to the bus interface devices. The cards within the system determine whether they are being addressed and translate the subsequent data stream into 1149.1 bus operations, returning the data received from the board for analysis.

For both the processor-based and autonomous embedded test controller-based systems, the data transferred between the controlling workstation and the board under test is usually command-based. The set of commands that are supported are specific to each board but will usually support both a "run everything" mode and a mode where specific tests can be executed. The processor or embedded test controller receives the command over the system's functional communication bus and initiates the test resequencing or pattern generation function required by the command.

The information returned by the on-card processor/embedded test controller is implementation specific, but, at minimum, indicates that the command either passed or failed. For completeness, the commands implemented on the card should include basic 1149.1 operations so that debug vectors generated external to the board under test can still be applied.

Pattern Resequencing

After a test pattern has been generated for a board, the pattern resequencing problem is to target the test pattern to a board within the system. For most 1149.1 backplane extensions, this entails adding addressing information to the test pattern. Most 1149.1 test systems, such as ASSET or Intellitech, support many of the extensions available. Therefore, from a user perspective, there is no difference in how tests are applied within a system compared to card edge access. The tool requires that the target address is specified and performs the pattern resequencing function directly.

For systems that use processor-based or embedded test controller access, the board-level test patterns must be converted into a (binary) pattern understood by the software or hardware using the data. Because the pattern is stored directly on card, there is no addressing component to be considered. Also, the format of the stored data is specific to the application using it, because there is no standard binary representation of the 1149.1 data. Using the 1149.1 data format specified within the 1149.5 standard is strongly recommended.

Test Pattern Generation

A critical factor to be considered when generating test patterns is the effect of those patterns on other circuit cards within the system. Tests run on a circuit card within the system must never toggle any backplane nets, unless a backplane net has to be driven to a constant "safe" value. (Testing of the backplane signals is performed at the next level of hierarchy.). Similarly, if a processor-based embedded test approach is used, all patterns generated for the circuit card must accommodate constraints imposed by the necessity of keeping the processor running.

Revision Control

The difficulties encountered when performing board-level tests in-system encompass both technical and data management issues. The technical issues pertaining to non-interference and recovering from a structural test have already been addressed. However, the issue regarding revision control of test patterns is of equal importance. When implementing an architecture to support access to board-level tests within the system, the capability will be available throughout the life cycle of the product and encompasses verification test, factory test, field test, and field return test. For every circuit card type and revision, a database of the structural tests must be maintained. This database must be accessible to every operator performing the test and must contain the suite of tests for every circuit card still in service. The complexity of maintaining this database is further compounded if a single revision of a circuit card has multiple configurations, either by using daughter cards or different memory configurations.

If tests are applied from an external source, using either the card-edge 1149.1 or 1149.1 backplane extension, the operator must first determine the type, revision and configuration of the board being tested. Short of pulling the card from the system and performing the task manually, the availability of a board ID (as described within the 1149.5 standard) that can be read from the board will help automate the task. After the board parameters have been determined, the vectors required for the specific board can be retrieved and applied. The failure analysis might also be specific to the revision of board under test, requiring that the results of the test be compared to a fault dictionary that is specific to that board revision.

Systems that implement their embedded test controllers as processor functions, or by using autonomous embedded test controllers, require on-card storage of test vectors. By storing vectors on card, the test data is always available. The revision control issue becomes one of a manufacturing process where any changes made to the circuit card will be reflected in the test vectors stored on card. Whereas this is easily facilitated when all repair or rework is done within the factory, in-field upgrades will require a change in the test data contents. This is an issue particularly when daughter cards are used and a technician can easily change the test configuration of a board. Another consideration is that these cards typically have two sets of test data associated with them: one for testing the card from the local processor and the other when testing the card from an external source in which the processor is included as part of the test. To extend the idea of using memory on the card to store test data, both patterns should be stored on card while providing access to the off-card test data in a manner that is autonomous to the processor operation.

System-Level Test Access

In the final level of hierarchy, the objective is to control the entire suite of tests available within the system. At this level, all inter- and intra-board functions are initiated, either from a single test connector external to the system, or remotely using network connections to the system. The infrastructure needed at this level is a function of the capabilities distributed across all cards within the system and the access level required for debug and diagnosis. The only test that is unique to an entire system is the testing of the backplane interconnect, during which the entire system function is disabled. This is in contrast to board-specific tests, where it is possible to maintain normal system operation while testing of an individual board is being conducted.

For the first time in the hierarchical test strategy, the design engineer must evaluate which test and maintenance functions need to be available within the system, and whether these functions need to be integrated as part of the overall system. The test and maintenance strategy of the system is based on two common criteria. The first is to determine the functions required to monitor system operation and perform routine *online* test and maintenance operations. The second is to determine the functions required to maintain system operation in the presence of faults and diagnose the cause of the failure without further degrading system performance. The mapping of these requirements onto the physical buses or communication paths within a system is then considered to ensure that test access is available wherever required.

In the previous section, the concept of a distributed test architecture was introduced, where cards within the system have the ability to test themselves, either under the direction of a central test controller or through an event that occurs on the circuit card itself. This architecture parallels the functional operation of a card within a system where a card is responsible for monitoring its activity, and taking any corrective action required to maintain system functionality. If a severe failure is detected, the faulty board signals the system that it can no longer perform its normal function and proceeds with diagnostics locally in order to collect as much information about the failure as possible. The system communicates with the board using high-level commands to decommission the card and aid in the diagnostic process.

This does not mean that the capabilities provided by 1149.1 backplane access are not required. In fact, for some failure modes, and for diagnostics, direct access to each card's 1149.1 bus is a necessity. Access can be provided from a central point within the system, which can be enhanced to allow network access to the 1149.1 subsystem using the infrastructure in place for normal test and maintenance purposes.

Hardware Requirements

Using 1149.1 backplane extension is the simplest implementation that allows access to embedded test and backplane interconnect test capabilities. In this strategy all test control and pattern generation functions remain external to the system under test. As such, these functions cannot be integrated as part of the overall system test and maintenance strategy.

In-system access to an 1149.1 backplane extension can be provided by using a backplane test bus master. This device provides processor access to the backplane test bus, allowing software to utilize any testing capabilities controlled by the backplane. In this implementation, the pattern generation and test control functions reside on a central card within the system.

Finally, if no test bus exists, access to a board's 1149.1 bus can be provided using functional communication paths that already exists within the system. This approach requires that each card within the system implements the bridging function between the communication bus, and the 1149.1 bus, or that each card implements the pattern generation functions required to initiate a test based on a command sent over the functional communication paths.

Test Control and Pattern Resequencing

The test control and pattern resequencing functions at the system level become a function of the overall system test strategy. If only simple 1149.1 access is required, the test control and test pattern sequencing function exist outside the system under test, and communicate with the system under test across an 1149.1 backplane extension. The strategy is identical to that required to test an individual board within the system, and assumes that the test and diagnostic capabilities available using embedded test are not going to be integrated as part of the system software.

The following questions must be answered to determine whether a simple 1149.1 backplane extension is sufficient to support the test control and pattern resequencing functions within the system:

● Do any cards within the system provide processor access to their 1149.1 bus?

● Are the embedded tests going to be used as part of the system diagnostic software, or combined with other test and maintenance functions?

● Are there physical, or architectural reasons a dedicated test bus cannot be implemented?

● Is more than 1149.1 access required over the backplane test bus?

If the answer to any of these questions is *yes*, the ability to access the 1149.1 bus on a board must be integrated with other test and maintenance activities. The test control and test pattern sequencing functions must also be integrated with the overall system test and maintenance environment, and might consist of a combination of strategies used at the board and in-system test access stages.

Test Pattern Generation

ATPG support is required to generate the backplane interconnect test, regardless of the infrastructure in place to apply the test pattern to the boards. There is no hierarchical interconnect test generation capability available commercially to simplify the task of generating a backplane interconnect test. The current methodology requires that the backplane netlist and board netlists be flattened so that the system looks like a single board. When this representation of the system and its backplane is available, the pattern is generated assuming a single scan-chain board implementation. The resulting pattern is manually partitioned so that the data is associated with a card within the system. These patterns are then interleaved with addressing information, and the pattern is applied to each card in the system.

The algorithm for a backplane interconnect test is slightly different than that used on a board. Since the boards within the system do not go through the *updateDR* state synchronously, the pattern generated must start with all devices driving the backplane tristated. This ensures that when each card outputs its data, no contention with other cards within the system occurs. With conventional 1149.1, the response data is captured and shifted one board at a time. When a board performs its capture, the output response is shifted out while the same input pattern is shifted in to maintain the data on the backplane. This process is repeated for the number of boards in the system. LogicVision has devised a way to overcome this limitation of the standard by providing a synchronization mechanism between chips of the different boards. The test is performed using system clocks allowing to test the interconnect at speed.

The process is cumbersome to manage, and is specific to a single system configuration. If a card is installed in a different slot, or the revision of a card is different than used when the backplane interconnect test was generated, the test becomes invalid. Another test pattern must then be generated and maintained. A truly hierarchical test generation tool is required to improve the management of a backplane interconnect test.

One possibility to manage the complexity of the backplane interconnect test problem is to have each card within the system responsible for generating the stimulus to be applied on the backplane based on a predefined algorithm, and high-level commands. A series of commands, such as "apply pattern 1" would be sent to each card within the system, who would generate the correct backplane stimuli and compare for proper backplane data, independent of revision or slot ID of the card involved in the test.

Sample Embedded Test Flows

The following section gives examples of using LogicVision technologies at the system level. The examples are based on specific system infrastructures. However, because all technologies are controlled from the 1149.1 bus, they can easily be adapted to control any LogicVision technology.

Example 1: LogicVision/ASSET InterTech Flow

Before any test data can be generated for any board or unit, it is necessary to create a description of each of the boards to be tested in a form that is used by the ASSET system. These descriptions, or ASSET database (ADB) files, contain information about the board's scan chain, the order of the chips on that chain, the various TAP instructions supported within each of the chips.

An ADB file for a given board design is created using one of the two flows depicted in Figure 8-8 on page 213.

The flow in Figure 8-8 *(a)* is the simplest. As shown, the ASSET tool box is used to create the ADB file based on the Boundary Scan Description Language (BSDL) files for each of the ASICs and the Hierarchical Scan Description Language (HSDL) file for the board itself. The BSDL file for each ASIC whose boundary scan chain was created by LogicVision is output by the top-level integration tool. The board HSDL file can be created manually or using the ASSET software. A board HSDL file is relatively simple as it essentially lists the chips that are on the board, and their order within the board boundary scan chain. For a complete description of HSDL files, refer to the *Hierarchical Scan Description Language (HSDL) Reference* manual from ASSET.

The disadvantage of the flow in Figure 8-8 *(a)* is that it only supports go/no-go embedded testing. This is because the BSDL files do not contain information about the structure of the embedded test controllers. To support some level of embedded test-related diagnosis, the flow in Figure 8-8 *(b)* is required. As shown, the BSDL files are replaced by chip-level HSDL files. LogicVision's extractor tool automatically creates an HSDL file which takes into account all of the embedded test controllers which have been embedded into the given ASIC. However, ASSET does not currently support all of the HSDL constructs used by LogicVision, and therefore some filtering of these files is necessary. This filtering is rather straightforward.

Figure 8-8 Asset Database Creation Flow

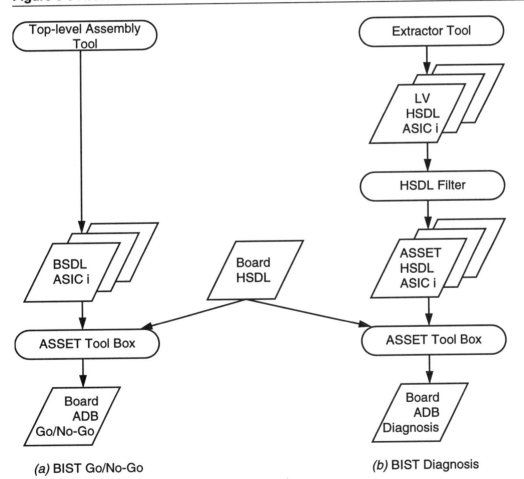

(a) BIST Go/No-Go

(b) BIST Diagnosis

The test development flow required to allow an ASSET system to run chip-level embedded test resources is shown in Figure 8-9. LogicVision's Verify tools create, respectively, the test data needed to run any logic BIST or memory BIST controller on a given ASIC. This test data is provided in SVF files. An SVF file consists essentially of TAP instruction (SIR) or data (SDR) scan operations.

The generated SVF files could be applied directly to the appropriate chips on the board. However, SVF does not provide any way of interpreting results other than pass/fail. It also does not contain any messaging capabilities. As a result, it is necessary to transform (as well as merge for convenience) the generated SVF files into a board-level ASSET macro file. The *svf2aml* utility performs this transformation.

Figure 8-9 Flow to Access Chip-Level Embedded Test at the Board Level Using ASSET

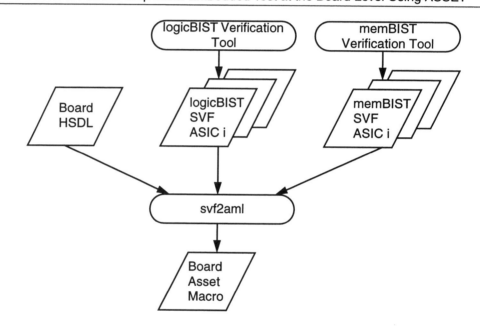

Note that if only diagnosis down to the failing embedded test controller is ever required, then a single ASSET macro for each board design is sufficient for both go/no-go testing and board diagnosis.

Example 2: LogicVision/Intellitech Flow

The top-level description of the board scan chain is contained within a device chain description (DCD) file. This file contains information pertaining to the ordering of devices on the scan chain, the association of the device reference (U1) to device type, and a reference to the device's BSDL file. The flow required to create the board-level database is shown in Figure 8-10 on page 215.

The BSDL files generated by *tapbistAssemble* are referenced within the board DCD file. The DCD file is converted into a board-level database, or DBS file, using the *File – Database – Create* menu option within Eclipse.

Figure 8-10 LogicVision/Intellitech Flow

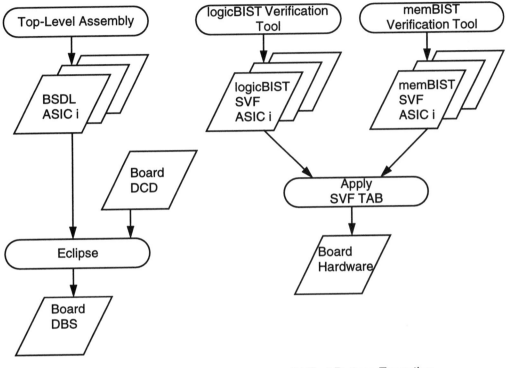

(a) Database Generation Flow   *(b)* Test Pattern Execution

The LogicVision Verify tool specific to an embedded test controller is run to create the ASIC-level SVF file. These files are referenced within the Intellitech Test application control window. To open the test application window, from the *Vectors – Test Application* menu, select the SVF files to be run. Select the Compile and Apply checkboxes to apply the SVF file to the unit under test. Failure results are reported as miscompares within the SVF pattern. Currently, failure diagnosis is performed manually by determining the position of the failing bits and mapping this to the data register within which the failing data was captured. The failure reporting and analysis will be enhanced in the future.

References

[1] *Hierarchical Scan Description Language (HSDL) Reference* manual from ASSET InterTech, Inc.

[2] H. Hulvershorn, P. Soong and S. Adham, "Linking Diagnostic Software to Hardware Self-Test in Telecom System," in *Proceedings of the 1995 International Test Conference*, pp. 986-1001.

[3] "IEEE Standard Module Test and Maintenance (MTM) Bus Protocol," IEEE Standard P1149.5-1995, January 1996.

[4] K. Parker, *The Boundary-Scan Handbook*. Kluwer Academic Press, 1992.

Glossary

This glossary defines terms and acronyms used in this book.

access   Mechanisms by which signals can be propagated to and from a core, from either encapsulating circuitry or from the primary inputs and outputs of the system chip. There are several types of access that are important to define. These are:

- *Functional access*—mechanism for propagating and observing signals during functional or *mission* mode.

- *Test access*—mechanism for propagating and observing signals during test mode. In the context of cores, it frequently involves delivering test data to and from a core while ignoring or isolating other cores that are connected logically during functional or mission mode.

Analyze tools   Tools that verify the testability of an original design and provide information to help improve testability by identifying redundant faults or nodes that are resistant to test. The signatureAnalyze tool can be used to create scan patterns (ATPG).

ASI   At-speed interconnect test.

ASIC   Application-Specific Integrated Circuit. This is the target for all embedded test-generated logic; also known as *the design*.

Assemble tools   Tools that automate the process of incorporating and connecting the design objects created by the Generate tools into the design.

asynchronous testpoints   Testpoints that use a test mode (TM pin) to enable or disable their effect. These testpoints do not require a clock to reset them.

ATE   Automatic Test Equipment.

ATPG   Automatic Test Pattern Generation.

at-speed test   Testing chip designs at actual clock speeds, rather than slower external tester speeds. LogicVision's at-speed test is scalable for clock speeds of 100 MHz and greater.

automation tools   Software that automates the process of adding embedded ATE design objects into the target design, configuring design objects based on the target design and the input, instantiating and connecting design objects, and providing additional support. This added support includes rules checking, scan insertion, testpoint insertion, signature calculation, fault coverage analysis, and test and diagnostic pattern generation.

bed-of-nails test technique   A method of testing that uses a fixture containing a bed-of-nails to access individual devices on the board. Contact is made through test lands laid into the copper interconnect or other convenient contact points. Fundamentally, the in-circuit, bed-of-nails technique relies on physical access to devices on the board.

BIST   Built-In Self-Test. A hierarchical DFT strategy that reduces the need for external test. With BIST, a small tester, complete with a pattern generator, is brought onto the chip, enabling the chip to test itself.

black-box model   A scan model that contains only a module declaration and input/output declarations.

boundary register   A boundary-scan serial shift register of boundary-scan cells that is used to observe and control the pins of a chip; also known as *BScan chain*.

boundary scan   The concept of a serial shift register around the boundary of the device; basis for the IEEE 1149.1 standard proposed by JTAG.

boundary-scan cells   Part of the 1149.1 TAP standard, boundary scan adds scannable logic cells to the pins of a chip. These BScan cells provide two test modes:

- *Internal Test Mode*—enables control and observation of the internal functions of the chip via the boundary-scan cells.

- *External Test Mode*—enables control and observation of the pins of the chip via the boundary-scan cells.

BScan chain   Boundary register.

BSDL   Boundary Scan Description Language. The standard description language for boundary-scan devices that comply with the IEEE standard 1149.1.

bus contention   Two bus drivers simultaneously active and in opposite states.

CA   Cellular Automaton. A type of random pattern generator.

capacitive load   Electrical property proportional to the amount of logic connected to a point in the circuit. A higher capacitance results in longer signal propagation delays.

capture operation   A test operation that loads into boundary-scan input cells those signal values that arrive at device input pins and loads into boundary-scan output cells those signal values that pass from core logic to device output pins. For internal scan and logic BIST, the capture operation occurs during the clock cycle that immediately follows the shifting of a test pattern into the scan chains.

chip pin   The actual connection to the outside of the chip. Also referred to as *pin*.

clockBscan   Test clock signal for boundary scan.

clock domain   The combination of a clock distribution network and all flip-flops that are driven by that network. Delays of clock distribution networks vary considerably, depending upon the cells that make up the network itself and upon the capacitive load that must be driven by the network.

collar   A hardware design object used to isolate a functional block of a design. This isolation allows the block to be tested without interference from its surrounding logic. A collar can also be used to provide controllability and observability of the surrounding logic.

combinational logic   Logic that does not contain any flip-flops, latches, or memory components.

control testpoints   Testpoints used to control the value of a circuit node to a logic one or zero value. The node is forced to the desired value during test mode. In normal operation, control testpoints are disabled so as not to affect the intended operation of the circuit.

core   A *design block* that is targeted for use in multiple designs. In this document, any design block that is fitted with a *test collar* or *test wrapper* is called a *core*. Cores can be designed by outside sources or by the user.

core block   One of the two blocks that the chip-level design contains, that identifies the top-level core design. The other block—the Pads module— includes the logicBIST controller, the 1149.1 TAP, the clock prescaler, the scan-chain router, and boundary scan and pad cells.

core integrator Person who puts together one or more cores with their own designs to create a system chip.

core provider Vendor of the intellectual property (IP) or predesigned function that can be reused by customers in their system.

core terminal See *terminal*.

core test Test that is applied to the core as a stand-alone device.

design block A partitioned section of a design. Design blocks are also commonly called modules or entities. A design block can contain one or more *subblocks*.

design objects User-configurable IP (intellectual property) design objects, delivered as RTL soft cores. LogicVision design objects include configurations for analog-to-digital BIST, at-speed interconnect, legacy core test collars, external memory BIST, logic BIST, memory BIST, phase-locked loop BIST, and 1149.1 TAP integration.

LogicVision design objects implement pattern generators, either random or algorithmic, results compression, collection of diagnostic data, and precision timing for at-speed test. Also referred to as *hardware objects*.

DFT Design for Test. The introduction into a design of features that make the design easier to test. These features enable test development automation by removing complex, sequential behavior and by implementing test infrastructure. This method also provides access (control and observation) to otherwise *hidden* internal states of the circuit.

DIN Signal that supplies input to boundary-scan cells.

DOUT Signal that supplies output from boundary-scan cells.

DR Test Data Register.

DRAM Dynamic Random Access Memory. A type of semiconductor memory in which the presence or absence of a capacitive charge represents the state (0 or 1) of a binary storage element. The charge must be periodically refreshed.

DUT Device Under Test.

effective scan-chain length In at-speed test, the number of flip-flops in a scan chain multiplied by the *ratio* of the highest system clock frequency to the clock frequency of the scan chain.

embedded logicBIST controller A logicBIST controller that can be optionally included in a test-collared embedded core.

EMI Electro-magnetic interference.

fanout stem A net that is connected to an input of more than one logic gate.

flip-flop A logical circuit that has two stable states—on and off—and a clock that enables state transitions. A basic logic circuit component.

FSM Finite State Machine.

functional test A test process that checks some aspect of the expected design behavior, for example, performance at-speed. Functional testing verifies the input-to-output behavior of the circuit.

Generate tools Tools that create synthesizable Verilog or VHDL RTL descriptions of embedded ATE components (controllers, boundary-scan cells, clock prescaler, scan-chain router, test access port) that provide at-speed test of various blocks in a design.

Each of the Generate tools creates synthesis scripts that accompany the RTL. These scripts create technology-specific, gate-level descriptions of design objects. The scripts can be modified to add design-specific constraints.

glue logic   Miscellaneous logic that is not application-related but necessary for connecting elements of the design.

HDL   Hardware Description Language.

hierarchical logic test   An embedded test scheme that consists of global (chip-level) logic test infrastructure and *test-collared cores* that contain *embedded logicBIST controllers*. The chip-level test hardware includes scan test or logic BIST of the interconnect between the test-collared cores and any UDL. The test manager is an IEEE 1149.1 TAP.

HSDL   Hierarchical Scan Description Language.

IC   Integrated circuit.

IDDQ   A test that measures the leakage current of an integrated circuit in the quiescent state in an attempt to verify the structural integrity of the circuit.

input observation   Process of observing data coming to the core from outside or from the interconnect.

Instruction Register   The Instruction Register (IR) selects the test to perform and/or the test data register to access, and also allows shifting out of the status information. This register is part of the TAP.

IP   Intellectual Property. The circuits produced by LogicVision tools are intellectual property that is proprietary to LogicVision.

IR   Instruction Register.

isolation   See *input control*, *input constraint*, *output control,* and *output constraint.* These definitions are related to isolation in that isolation encompasses these operations.

JTAG   Joint Test Action Group. This organization devised the IEEE 1149.1 standard test architecture that is widely used in the test of boards and systems.

least significant bit   The lowest order or the right-most bit of a register. The least significant bit corresponds to bit number 0. Also known as *LSB.*

legacy core   A core that cannot be significantly altered to introduce DFT features. Legacy cores normally consist of a completed transistor-level layout database but are sometimes provided as an RTL description or a gate-level netlist. The defining constraint is that a legacy core cannot be altered to enhance its testability.

Legacy Core Test option   Capability that surrounds a legacy core with an isolation-and-access test collar and connects this collar to chip pins. These connections allow the designer to directly apply functional test vectors to the core from the chip pins.

LFSR   Linear Feedback Shift Register. A hardware random pattern generator.

Logic BIST configuration   In the logic BIST configuration *(-mode icBIST),* the scan chains in the core block and the boundary-scan chain (which might consist of several segments) are individually linked to pseudo-random pattern generator (PRPG) and multiple-input signature register (MISR) ports. See also *scan-chain configurations* on page 222, *multi-chain configurations* on page 221, and *scan-through-TAP mode* on page 223.

logicBIST controller   A block of logicBIST circuitry that consists of a pseudo-random pattern generator (PRPG), a multiple-input signature register (MISR), and the logic required to perform various control functions.

LSB   See *least significant bit.*

MISR Multi-Input Linear Feedback Signature Register. An MISR is typically used by LogicVision tools as a signature analyzer.

most significant bit The highest order or left-most bit of a register. Also known as *MSB*.

MSB Most significant bit.

multi-chain configuration The multi-chain configuration (*-mode Multi*) is similar to the scan-through-TAP configuration in the sense that all of the scan flip-flops in the design are included in the configuration. In the case of the multi-chain configuration, the scan flip-flops are segmented into several parallel scan chains. The designer specifies the number of scan chains while completing the Top Flow. See also *scan-chain configurations* on page 222, *scan-through-TAP mode* on page 223, and *logic BIST configuration* on page 220.

multi-chain configuration A path that requires more than one clock cycle for the data to traverse the path. In the design of certain types of functionality, it is useful to incorporate multi-cycle paths into a design. For example, synchronous counters that use multi-cycle paths can run at a higher frequency versus counters constructed with the same technology that does not use multi-cycle paths.

mergeable core Cores that can be integrated into the DFT methodology of the system chip. a typical way to supply a mergeable core is to supply a behavioral description or RTL description. This kind of description can be more easily merged into the system chip and the entire chip can be synthesized with other cores with the ATPG scan chains merged into an overall test plan.

non-mergeable core Cores that cannot be integrated in the test methodology of the system chip. A typical non-mergeable core will have a physical design that will not allow the modification of a test methodology. It becomes necessary to modify the system chip design to port the core test to the core.

non-shared wrapper (cells) Wrapper style that does not share hardware with functionality. This style allows test operations to occur during functional operation.

normal mission mode The functional operating mode of the device under test (DUT).

observation testpoints Testpoints used with difficult-to-observe circuit nodes. The difficult-to-observe node is sampled during scan-capture mode; the captured value of the difficult-to-observe node is shifted out of the scan chain.

P1500 IEEE standards project group, with associated working groups, charged with developing standard test access and test protocol methods for testing embedded virtual components.

parallel-compare operation In logic simulation, the action of comparing captured values to expected values directly on flip-flop outputs instead of shifting them out through the scan chains and comparing them on the scan-chain outputs.

parallel-load operation In logic simulation, the action of directly loading values into scan-chain flip-flops instead of shifting them in through the scan chain.

PGSL Pattern Generation Script Language. This language is internal to LogicVision and is used by the verification tools.

PI Primary Input.

PLL Phase-Locked Loop.

PO Primary Output.

post-synthesis tools Tools used during or after synthesis that enhance the testability of the synthesized gate-level description, prior to optimization, or verify the testability of the target design output from logic synthesis, and generate and sequence test patterns for this design.

Prep tools Tools for transforming certain design-related information, via the creation of input files, into data usable by the Generate and Assemble tools.

pre-synthesis tools   *Front-end tools* that accept high-level user parameters and generate testability-enhancing RTL design descriptions for compilation and optimization using any logic synthesis engine.

priority hold flops   Specially designed scan flip-flops that transmit data into multi-cycle paths.

PRPG   Pseudo Random Pattern Generator. The PRPG is used to generate test patterns for logic BIST.

RAM   Random-Access Memory. A type of semiconductor memory that stores digital information temporarily and can be changed as required. This memory type constitutes the basic (read/write) storage element in a computer.

redundant fault   A fault that cannot be detected with a test vector. If the logic contains built-in redundancy, observation flip-flops (testpoints) can be added to some of the circuit nodes to improve testability.

ROM   Read-Only Memory. A type of semiconductor memory that permanently stores information repeatedly used, such as tables of data and characters of electronic displays. Unlike RAM, ROM cannot be altered.

root   In the LogicVision Top Flow, the name of the top-level Verilog module of the design or the top-level VHDL entity declaration. Also referred to as *design module name* or *chip-level design.*

RP   Random Pattern.

RPR   Random Pattern Resistant.

RTL   Register Transfer Language. A design language format that is technology independent, can be Verilog or VHDL, and is compatible with Synopsys synthesis.

sample mode   Test configuration in which a flip-flop captures the value of a signal at the input to the boundary-scan cell (DIN).

scan-chain configurations   LogicVision's tools can implement up to three scan-chain-configurations, which include scan-through-TAP, multi-chain, and logic BIST. See also *logic BIST configuration* on page 220, *multi-chain configuration* on page 221, and *scan-through-TAP mode* on page 223.

Scan-DR   TAP controller state that applies the scanning process to each test data register. Comprises these individual controller states: Capture-DR, Shift-DR, Exit1-DR, Pause-DR, Exit2-DR, and Update-DR.

ScanFF   A flip-flop with two data inputs, usually a functional input and a scan input. While different types of scan flip-flops exist, most use a scan-enable input to determine which of the two data inputs to load.

Scan-in (SI)   Scan Data In. The serial input of a scan chain segment.

Scan-IR   TAP controller state that applies the scanning process to the instruction register. Comprises these individual controller states: Capture-IR, Shift-IR, Exit1-IR, Pause-IR, Exit2-IR, and Update-IR.

Scan mission mode   This mode occurs when TM=1 and SM=0. All flip-flops are clocked off the Scan clock (SCLK) or primary input clocks, and all flip-flops latch the normal mode D-input (rather than the Scan data). This mode is also referred to as capture mode.

Scan mode (SM)   This mode is a subset of the test mode. Scan mode reconfigures the flip-flops of the DUT (when SM=1) as a Scan register in order to shift in or out serial test patterns. The signal that causes the reconfiguration is also referred to as *Scan Enable.*

scan model library   A library of technology-dependent scan models. This library is necessary for rules checking.

Scan-out (SO)   Scan Data Out. The serial output of a scan chain segment.

scan-through-TAP mode   The *scan-through-TAP* configuration serially connects the scan flip-flops in the design into a single scan chain (shift register). This chain connects scan flip-flops in the core block, the logicBIST controller, the scan-chain router, the clock prescaler, the TAP, and boundary-scan cells.

The scan-through-TAP configuration is selected by issuing a command to the instruction register of the IEEE 1149.1 TAP. For scan-through-TAP, the TAP issues a set of control signals to the multiplexers of the scan-chain router block to create this configuration. See also *logic BIST configuration* on page 220, *multi-chain configuration* on page 221, and *scan-chain configurations* on page 222.

scannable circuit   A circuit that provides a mode where all the storage elements in the circuit are chained together into one or more shift registers, with external input and output. In this way, an arbitrary state can be shifted into the storage elements, the circuit can be operated normally for one (or several) clock cycle, and the resulting state of the storage elements can be shifted out. The use of scan enables high fault coverage because of the high degree of controllability and observability which it provides.

scanVerify   The Verify tool that generates a testbench that emulates the BIST operation. The pseudo-random vectors are applied to the scan chains and the results are captured and compared with the expected data.

script_bscan   A script that, when read into *dc_shell* with the include command, synthesizes all appropriate boundary-scan cell RTL files. For all of the rules-checking and logic simulations that employ boundary scan, the gate-level output from the *script_bscan* script is used.

script_testpoints   A script that, when read into *dc_shell* with the include command, synthesizes all appropriate testpoint RTL files. For all of the rules checking and logic simulations that employ testpoints, the gate-level output from the *script_testpoints* script is used.

SE   Scan-enable signal.

selectJtagSignal   Multiplexer-enable signal for boundary scan. Selects latch or input to boundary-scan cell.

shadow logic   User-defined logic accessible only from the input/output ports of virtual components (VCs). The VC is said to cast a shadow which potentially reduces the testability of the logic in that shadow. The addition of test access to the VC ports is said to cast light on or remove the shadow. If the UDL contains internal test access points (such as scan elements, which can act as both test control and test observation points) then the shadow logic includes only the logic between the BC output ports and the first level of test (control) access points in the UDL and the logic between the last level of test (observation) access points in the UDL and the VC input ports. Shadow logic that is partially accessible from other non-shadowed UDL via chip primary ports or test access points in the UDL is sometimes said to be partially shadowed as the ease of detecting faults on some logic nodes in the shadow logic might be unaffected because of this partial accessibility.

shared wrapper (cells)   Wrapper style that shares logic between test and functionality. Typically registered inputs and outputs that can be used in test mode to assert and capture test data.

shiftBscan   Multiplexer-enable signals that shifts or samples during boundary scan.

sink   Refers to test response evaluators. A sink can be either on chip (BIST) or off chip (external ATE). Usually connected to an embedded core via a Test Access Mechanism (TAM). Also referred to as *test sink*.

SM   Scan Mode.

SOC   System on a Chip. Devices created by using VDSM technology. With SOCs, the vast availability of gates is used to integrate analog circuitry, blocks of logic, memories, and processors on a single chip.

source   Used to refer to the test stimuli generator. A source can be either on chip (BIST) or off chip (external ATE). Usually connected to an embedded core via a TAM.

SRAM   Static RAM. A type of semiconductor memory in which a pair of flip-flops is cross-coupled to hold a binary state as long as power is applied.

structural test   A test process based on a pattern set that has a known defect or fault coverage. This testing is used for manufacturing. Structural testing involves strategies to verify the individual elements of the design: logic gates, transistors, and interconnects.

subblock   A subdivision within a *design block*. Sample subblocks are *cores*, memories, and UDL.

SVF   Serial Vector Format. A board-level test pattern format used with IEEE 1149.1 board test software. SVF was jointly developed by ASSET InterTech and Teradyne.

synchronous testpoints   Testpoints that require one or more clocks to reset them.

system chip   A system integrated on a chip. Included are one or more cores with UDL integrated by the core user or system integrator.

TAP   Abbreviation for the five-pin, IEEE 1149.1 Test Access Port.

The IEEE 1149.1 TAP defines a chip-level test structure that is used to implement a simplified board and system-level test strategy.External test processors control and communicate with various test features built into a design through the TAP. The TAP is used to initialize BIST operations, start BIST, monitor status, and output results.

The five pins of the TAP are TDI, TDO, TCK, TMS, and TRST.

TCK   Dedicated Test Clock. This mandatory terminal of the TAP is driven by the clock that controls the synchronous operation of the TAP.

TDI   Test Data-In. Serial test data shifted in with a default value of 1. A mandatory terminal of the TAP.

TDO   Test Data-Out. Serial test data shifted out with a default value of Z. This mandatory terminal of the TAP is only active during a shift operation.

TDR   Test Data Register.

terminal   Connections at the input and output of a core. The term *node* is sometimes used but *terminal* is the preferred term.

test access mechanism (TAM)   Logic that provides data transport between sources of test stimuli and embedded cores and between embedded cores and test response evaluators or sinks. The core integrator adds TAM features to deliver data to the core wrappers or cores. The TAM delivers test data between cores and various pins of the chip. The TAM also delivers dynamic controls the cores or wrappers.

testbenches   Output from the Verify tools that initializes and executes tests then returns test completion values (DONE flags, pass/fail signals). These testbenches also provide diagnostic information useful for debugging. These testbenches provide input to simulation tools.

test collar A *test wrapper* used to isolate a functional block of a design. This isolation allows a *core* or other design entity to be tested without interference from its surrounding logic. A test collar can also be used to provide controllability and observability of the surrounding logic. Sample design entities that might employ test collars include *cores*, memories, and non-scannable logic.

test-collared core A core that has been fitted with a *test collar* or *test wrapper*. A test-collared core can also be modified to embed structural test infrastructure such as scan and logic BIST. A test-collared core can contain subblocks such as test-collared *legacy cores* and memories. Also referred to as *collared core*.

test-collared core assembly A design block containing a *core*, a *test collar* or *wrapper*, and, in some cases, a *logicBIST controller*.

Test Data Register User-defined scan chains that are connected to the IEEE 1149.1 TAP.

test integrator Person focused on the testability of the system chip, and with collecting, preparing, and verifying information needed to generate the test-program that is used during the manufacture of the system chip.

test mode (TM) A mode in which specific signals and configurations are enabled to test the DUT.

Test mode signals are active for the full duration of the test. In this mode (TM=1), all scannable flip-flops must be clocked off SCLK, the internal Scan clock when a test access port (TAP) is used. Otherwise, all flip-flops must be clocked from primary inputs.

Scan design rules are checked only when TM is active. This allows for the reconfiguration of the ASIC circuitry.

test period Any one of three test frequencies assigned to a clock domain in the top-level core block. Test frequencies can be f (highest), f/2, and f/4. scanGenerate uses these frequency assignments scanGenerate to optimize the scan-chain partitioning for a minimum test time.

testpoints Small blocks of logic that can be inserted into a netlist to improve controllability or observability of difficult-to-test nets.

test sink See *sink*.

test source The source of test data.

test wrapper See *test collar*.

test wrapper cell Element of the wrapper associated with a single signal.

TM Test Mode.

TMS Test Mode Select. Serial input control signal with a default value of 1. A mandatory terminal of the TAP.

top-most design block Represents the system chip top-level logic. Often contained within a single design entity but can be provided as multiple objects. Also referred to as *top-level core block*.

TPI Testpoint Insertion.

transparent mode A mode for transparent latches in which the gate of the latch is always active; data flushes directly through to the output.

TRST Test Reset. Asynchronous TAP controller reset with a default value of low. This optional terminal of the TAP usually has an active low state.

TTL Transistor-to-Transistor Logic.

update operation   A test operation that passes signal values already present in the BScan input cells into the core logic, instead of passing the values on chip pins. This operation also passes signal values already present in the BScan output cells to the chip output pins, instead of passing the output values generated by the core logic.

updateBscan   Latch-enable signal for boundary-scan cells.

UDC   Up/Down Counter.

UDP   User-Defined Primitive.

userBits   Extra bits that are manually added to the TAP instruction register.

userDRBits   Extra bits that are manually added to the TAP data register.

User-Defined Logic (UDL)   In general, UDL is all logic that is designed by the system-chip integrator that is not part of a core.

VDSM   very deep submicron. A technology for designing ICs with high clock speeds (100 MHz or greater) and millions of gates. This technology enables increased functionality, higher chip performance, and decreased costs. Typically, VDSM designs are manufactured with CMOS processes with gate lengths less than or equal to 0.25 microns.

VHDL   VHSIC Hardware Description Language.

VHSIC   Very High-Speed Integrated Circuit.

Verify tools   The Verify tools create testbenches to verify the proper operation of the scan chains or embedded ATE blocks that LogicVision Generate tools add to user designs, as controlled through the IEEE 1149.1 TAP or a direct pin interface.

virtual component (VC)   Another term for core.

WGL   Waveform Generator Language. A TSSI-proprietary pattern format.

WRP   Weighted Random Patterns.

wrapper   Syntax that describes software format. Wrappers have the format:

```
wrapperName{
    property1:value;
    property2:value;
    propertyN:value
}
```

where *wrapperName* is the name of a wrapper and *property* represents an attribute associated with the wrapper.

Index

Q